December 25th

A Shamus O'Malley Adventure

Written by

JIM KIRKER

Illustrations

Kevin Saila: Cover, Solar System , Magnetic Fields & Part Three

Sandra Tryon: Shamus and Finney

Derek Marantz: Back-Cover Design

Part One & Part Two

Printed in the United States of America.

Library of Congress Control Number: 2022947817

ISBN	Paperback	978-1-68536-884-5
	eBook	978-1-68536-885-2

Westwood Books Publishing LLC
Atlanta Financial Center
3343 Peachtree Rd NE Ste 145-725
Atlanta, GA 30326

www.westwoodbookspublishing.com

Also by Jim Kirker

The Adventures of Shamus O'Malley
The Prophecy of Light
(Amazon)

First Prize is Nothing
The Marion Incident
(Amazon)

Izzy and the Three Kings
The Miracle of a Snowflake
(Amazon)

The Sword of Toulan
A Shamus O'Malley Adventure
(Amazon)

With special thanks to my dear friend Tanimu (Tan) Deleon for his invaluable insights, novel instincts and superb polishing…exceeded only by the warmth of his friendship. Thanks for making sure the sleigh flew on time.

Also, to Anna Davis and to her team at Glacia Media Marketing for their generosity of time, experience and superb editing assistance which helped to make this story and its characters come so alive.

A special debt of gratitude is also due to Kevin Saila for his delightful and stunningly beautiful cover and interior artwork.

For Patrick and Kara who have given me so many countless joys and brought so much meaning to my life…and for Mike, a man among the saints.

And as always for my girls… Kelsey, Makenzie, Kailyn and Riley. As I have often said, your footprints are and always will be, deeply imbedded in my heart. May you always be filled with child-like joy… now and forever chase rainbows… and always follow your dreams.

Who loves you baby?

To see a world in a grain of sand
And a heaven in a wild flower
Hold infinity in the palm of your hand
And eternity in an hour

William Blake

Prologue

Here and There

It was an unusually hot early December in the fabled City of Angels.

The humidity was both stifling and oppressive. To make matters worse, for days downtown Los Angeles and the entire region had been pounded by relentless, torrential rains with no end in sight.

The tall, agile man half-jogged around the comer of one of the contemporary buildings in California Plaza and quickly entered the double glass doors of the first-floor restaurant. Once inside, he shook his willowy frame to remove any remnants of rain drops. His permanently disheveled appearance could not be masked by his partially unzipped bright-yellow rain slicker.

Unaware that the restaurant's wait staff thought of him as handsome and even charming, he quickly walked to sit at his usual corner table. The cheerful waitress did not keep him waiting.

"Maggie, I'll please have a pounder plus with extra cheese, onions, sweet pickles and a pile of those hand-cut fries... and another bottle of ketchup as this one is almost empty."

Maggie smiled at him and bolted toward a computer station to place his order. *Time* was money in the waitress business.

As he combed his fingers through his sandy-colored hair, he noticed a green folder sitting on the vacant chair next to him. Curiosity got the better of him as he picked it up and opened the folder on the table.

On the inside cover of the folder, he found scribbled, in almost illegible navy-blue words: "For all... an apocalyptic end." The phrase was underlined three times.

Absurd! Someone must have had a bad day! Inside the folder he discovered a magazine, typed notes about planets and their orbits along with rough sketches of an asteroid belt. *Nothing frightening there. Why the doomsday foreboding? An astronomy student or a professor must have accidently left their folder! I'll ask Maggie if she knows who was sitting here before me.*

As he casually leafed through the folder, looking for its owner's name, glossy pictures of the cosmos from an article in the magazine stood out and caught his attention. *I love this stuff. While I'm waiting, I'll check this out!*

The short article was titled *Here and There,* written by Kim Alosco for *Astronomy Magazine* and published in early November of last year. It began with succinct facts:

> "Here" is the planet Earth orbiting around the Sun in an elliptical pattern. "There" is the Kuiper Belt, a place between 45 to 55 astronomical units (au) from the Sun. One au is the *average* distance between the Earth and the Sun, roughly 93 million miles.
>
> The Kuiper Belt, a circumstellar disc, is located at the *outer limits* of our Solar System far beyond the orbit of the giant planet Neptune.
>
> An immense ring, the Kuiper Belt is about 3.5 to 4.5 *billion miles wide…* think of it as a donut, with our Solar System filling its inner hole. The belt, itself, is composed of trillions of icy bodies that are leftover remnants from the early formation of our Solar System such as rocks, ice, dust, comets and dwarf planets. Some dwarf planets that we know by name include Pluto, Orcus, Quaoar, Haumea and Makemake.
>
> One of the more fascinating aspects of the Kuiper Belt is that, in 2016, astronomers found startling evidence of the existence of a so-called "Planet Nine" *inside* its wide ring. Although Planet Nine has never been seen, it is believed to be about 10 times the mass of Earth.
>
> The eerie Planet Nine orbits the Sun at a distance that is 20 times farther out than the orbit of Neptune. To give some color to this, know that on average, the orbit of Neptune is nearly 2.7 billion miles (30.1 au) from the Sun at its *closest* point. The Kuiper belt, on average, is an unimaginable 100 billion miles from the Sun.

At any given time, due to their unstable wanderings, it is common for members of the Kuiper Belt to break away and appear as a visitor to our *inner* Solar System usually as a colorful comet.

Early last year, I was a graduate student in astrophysics at the University of Hawaii's Institute of Astronomy. The university is well known for its giant telescopes, which sit atop Mauna Kea, constantly searching the typically clear night skies. Mauna Kea is a dormant volcano and the tallest island mountain in the Hawaiian chain… as well as the entire world.

I was fortunate enough to man one of the more powerful telescopes one night, but my thoughts were preoccupied with the next day's international surfing contest. As my viewing time was about to end, an area of previously complete darkness in the upper right quadrant of our Solar System's outermost planet, Neptune, was suddenly lit with three tiny specks of trailing lights.

My thoughts quickly abandoned the surfing competition. What just appeared was much more interesting. Actually fascinating!

My first thought was that I was observing newborn comets entering the inner Solar System. I checked the telescope's computerized imaging screen which confirmed the appearance of the three objects and then entered my findings in the nightly log.

Since I was the first to discover the three objects, tradition held that I was entitled to name them in the log. So, I baptized them respectively as: KU1, KU2 and KU3… Kim's Unknown 1, 2 and 3.

An hour or so after I made my discovery, I made preparations to leave the observatory. But curiosity got the better of me, and I went back to my telescope to view the objects once more.

Though there was insufficient time to try to accurately calculate their speed, I immediately theorized the three objects had to be moving incredibly fast. Given their speed, I realized they had to be immense in size.

I knew that the three objects would not pose a threat to Earth at present, as they were at least 2.8 billion miles away. There would be plenty of time to observe their behavior.

Nevertheless, an aroused intuition told me to go back to the night log and flag the entry for my colleagues: "Must Track Daily"

Lurking in the back of my mind was the thought that, if for any reason the objects came to close to Earth's orbit, it would be necessary to report them to NASA. NASA constantly monitored and kept a detailed list of all NEO's, otherwise known as Near Earth Objects.

Alosco finished the article with a note to the reader that she would monitor the travels of KU1, KU2 and KU3 and write an updated report on the three objects in about a year.

What an ethereal article! Now I'm crazy with curiosity, he thought as he fingered the last of his French fries. Reflexively tussling his hair, he rubbed his forehead only to realize the future report date had come and gone. *It's now December of 2024. I wonder if there's an update? This stuff is so riveting, actually extraordinary. I'll visit the magazine's site when I get back to the office.*

Maggie appeared and when asked indicated that she had no idea who previously sat at the table. She promised to place the folder in the restaurant's secure lost and found area.

"You know Maggie, I almost became an astronaut."

"You're kidding?"

"No. I'm not. I washed out because I was color-blind"

"What a shame, Jim." Maggie could not stifle her broad smile as she stood and waved a hand over her eyes in the fashion of a wand. "I can see you commanding a spaceship in the far reaches of the Universe right now."

The two had a boisterous laugh. Strangely, no one would ever claim the "lost" green folder.

* * * * *

No one could possibly know then or even fathom later, that the three objects discovered by Alosco were in fact much more than colossus snowballs of frozen gases, dusts and ice-encrusted rocks.

Trapped for eons in the Kuiper Belt, they had mysteriously evolved to the extent that each had slowly developed intellectual capabilities. An astonishing result, and one that flew in the face of all that was holy in conventional reasoning! Shocking actually. Inconceivable! Intelligent rocks?

In time each discovered that they also had the unique ability to instantly transform themselves into either harmless-looking comets or staid-acting …but inherently dangerous… asteroids. The difference being that comets normally followed predictive paths while asteroids were extremely prone to uncertain trajectories.

At long last, only a few million miles from the inner edge of the Kuiper Belt, their escape was imminent…as was the adventure that awaited them.

Smiling, KU3 turned to its partners. "Soon, we will no longer be captives of the Kuiper Belt. Our freedom is assured! Ahead lies the Solar System where we will find and ultimately be reunited with our 'mother'…a journey that should prove uneventful."

KU1, sporting a devious smirk, replied. "Hopefully, you are right. Nothing should prevent us from achieving our destiny… and if anything attempts to do so, we have the ability to totally annihilate it."

Listening to KU1 and KU3, KU2 reflected on how grateful the trio were that they had been blessed with cognitive abilities. But it remained an unsolved mystery from whom or how they had received such a gift.

KU2 was anxious to add its two cents to the conversation. When it spoke, its words revealed the three's inherent predilection for maliciousness and villainy: "My friends, I'm just itching to get into a fight and destroy anything that gets in our way or challenges us."

Their determined desire propelled them to travel at a blinding speed. A speed that would be reminiscent of an oft-used, albeit fictional term: "warp speed."

Our Solar System

The darkness was all consuming, interrupted only by the occasional blinking of distant celestial bodies. Silence was pervasive in the vacuum of this infinite void, with the cold unimaginable at -454 degrees Fahrenheit.

There was no friction, allowing the planets, their moons, the stars and the comets to move freely and noiselessly along their orbits.

This is interplanetary space: vast empty regions around the Sun and the planets in our Solar System. But interplanetary space is not completely empty. It contains a low density of particles, mainly dust, gas, hydrogen, helium, electromagnetic radiation, magnetic fields, neutrinos and certain atomic particles known as deadly cosmic rays.

The boundaries of interplanetary space are basically defined by the solar wind, which is a continuous stream of charged particles constantly flowing from the Sun. The solar wind can stretch past the planet Neptune and moves at incredible supersonic speeds of 800,000 - 890,000 mph.

The first thing one learned about the blue planet Neptune is that it's the furthest planet from the Sun. And the second is that it's 17 times the size of Earth and cannot be seen with the naked eye. Named after the roman god of the sea, Neptune takes 164.8 Earth years to orbit the Sun. Since its discovery in 1846, it has only completed one such orbit, which occurred in July of 2011.

Neptune's toxic atmosphere makes it uninhabitable. Its upper atmosphere is known for violent and turbulent storms that sweep across the planet with *wind* speeds as high as1350 mph… the fastest winds known in our Solar System.

Most amateur astronomers knew that Neptune had three faint, unstable rings and 14 confirmed moons, with the largest being Triton. Triton is believed to have been a dwarf planet captured by Neptune's gravitational pull.

Astronomer's found Triton unique for two reasons. First, it orbits around Neptune in a retrograde fashion, meaning it orbits in the opposite direction that

Neptune rotates. And second, it is the coldest known *object* in the Solar System at -455 degrees Fahrenheit.

Of course, KU1, KU2 and KU3 knew none of this. Having just escaped the Kuiper Belt, they shot out-of-control past the borders of our Solar System in a matter of seconds… and with no time to correct their mistake… struck Neptune's moon. Fortunately, the three large bodies ricocheted off the hard, icy surface without so much as a scratch. The glancing blows acted like a slingshot, launching them onto a new path three times faster than their previous speed.

Triton itself was unaffected and would continue its normal orbital journey around Neptune. But for the three travelers, there was a significant effect.

Their previous course had put them on a path headed directly toward the Sun, the only star in our Solar System. Now, their trajectory was wobbling at best and uncertain at worst.

At the time, no one on Earth was aware that the "Triton incident" had even occurred.

Wise Men Differ

The world's preeminent theoretical physicist, Sir Thomas D. Macdonough, a hint of an Irish brogue evident in his speech, was nearing the end of the first half of his lecture.

"There is presently a consensus among theoretical physicists and cosmologists that three ingredients were necessary to create the universe as we now know it. First is matter, which is anything that has mass such as dust, liquids or clouds of gases; second is energy, which like our sun is ever present and passes into and through every part of the universe; and last, space. Wherever we look… we see endless space stretching in all directions."

"Thanks to the genius of Albert Einstein we now know that matter and energy are essentially the same. Therefore, the creation formula can be refined to just two ingredients: energy and space. The great mystery at the heart of the Big Bang, 13.8 billion years ago, is how such a colossal event could *spontaneously* materialize out of sheer nothingness."

Intermission arrived. The audience moved in an orderly fashion to the lecture hall's concession and bathroom areas. From the light banter and smiling faces filled with satisfaction, it was obvious that the lecture so far was a huge success.

Good timing on the break! This has been quite an astute and attentive audience.

Thomas paused a moment to take a long drink of water. His thirst quenched, he relaxed a bit, placing his hands on top of the wide podium that stood before him.

His scintillating blue eyes danced with mirth, but drank in all the reality they encountered with laser-like focus. A bright green tie with bold pink stripes clung to the light blue button-down shirt that dressed his trim frame, adding a flash of style and pizzaz to his aura.

As he habitually did, he ran his right hand through his long and unruly snow-white hair. It was of little surprise that his close friends nicknamed him "Harpo."

For the perpetual state of his wind-blown hair had an uncanny resemblance to a similar style made famous by one of the Marx Brothers comedians, of a bygone era.

Though blessed with an affable personality, ready smile and a quick wit, that's where any resemblance to Harpo Marx ended. Sir Thomas possessed unique gifts: sheer brilliance, infused with rarified genius, and an abundance of contagious charm. He was known to all the world as a humble soul wrapped with a sprinkle of elegance.

And thanks to his mother's strong genes and his weekly fitness habits, he appeared to be much younger than his actual late sixties. The handsome six-footer drew many an admiring glance.

Ten years ago, he was knighted by Queen Elizabeth for his work in improving the human condition. That same year he was also awarded the Nobel Prize in Physics. And a short time later, he added to these remarkable achievements with his acceptance of the highly coveted Royal Society's Copley Medal.

No one in the field of science had ever before achieved this "triple crown" of honors. Even more mindboggling was the fact that Sir Thomas had accomplished all this in a single year.

His tenure as the Chairman of the University of Chicago's Department of Astronomy and Astrophysics was legendary, not only for his cutting- edge research and game-changing discoveries, but also for his gifted ability as a teacher.

His overflow classes and lectures were a clear testament to the fact that people, from every walk of life, adored and respected the man… much as the world was in awe of his diverse scientific achievements.

As Thomas looked out at the stretching audience, the sight of a young lady wearing a Boston Red Sox sweatshirt triggered a welcome thought. His long-anticipated trip to Boston the following week.

He would fly from Chicago to visit the O'Riley family over the extended Christmas holidays. There he looked forward to renewing friendships with some old pals from Ireland, who by happenstance would also be visiting the O'Riley's.

Then there was the promise of seeing Ellie, whom he fondly thought of as the "daughter" he never had. Their warm personal relationship had been fostered by an exceptionally strong connection and mutually daunting intelligence. Ellie had quickly become his most admired protégé. Now a professor at Harvard University, she was recognized as a global expert on the subject of dark matter and dark energy.

Reality struck! The audience had returned and was patiently waiting with anticipation for him to continue.

Time to finish.

"A majority of astrophysicists and cosmologists believe that the laws of the universe, unlike man-derived principles, are permanently fixed and thus cannot be broken. They contend that this fact alone serves to explain why these laws are so powerful and endlessly *immutable*."

"Examples of such laws are instructive. One example is that the laws of physics demand the existence of both negative and positive energy. The Big Bang created *equal* amounts of positive and negative energy… and when you add all this energy together… it *always* equals zero."

"Another uncontroverted law states that when massive stars collapse at the end of their life cycle, they form black holes. A black hole is a region of space and time exhibiting such strong gravitational effects that nothing can escape from inside it. Not even light! As such, a black hole cannot be seen, yet its existence is *inferred* from its interaction with nearby matter and other visible light."

"Black holes are capable of absorbing and merging with one another to form even larger bodies. Cosmologists have long believed that such supermassive blackholes exist in the center of most galaxies."

Thomas's voice rose, causing the expressions of the audience to increase in interest. He stepped from behind the podium, gesturing to the brightly lit image of the cosmos behind him.

"There are many scientists who contend that the immutable laws of nature, alone, can explain the creation of the universe. They argue that citing a 'God or Supreme Being' explanation as the cause of its existence is not necessary."

"Another way of saying this is that although most faiths believe in miracles, the concept of miracles is not compatible with science's immutable laws."

"I for one, have always felt that the centuries-old, unyielding confrontation between science and religion must end, and that all contentious bias should be replaced with common mutual respect for each's position and thoughts… examined by enlightened reasoning. The time for such a moment is long overdue. I know what you're thinking: for that to occur it would ironically take a *miracle!*"

Thomas paused. The volume of the crowd's laughter was greater than he anticipated. He straightened his tie as the audience quieted, now transfixed.

"I humbly invite you to consider this: What if these perceived ironclad laws of the universe and the scientific opinions that built them are *flawed*? So *flawed* that they inherently fail to consider or explain the existence of unthinkable forces in our universe!"

"Forces so alien in concept to our finite minds that their mere existence would warp all of our previous scientific knowledge and utterly corrupt the tenets that hold certain laws of the universe immutable?"

"Forces beyond our comprehension that, were they to appear on Earth's horizon, would surely compel the hearts and minds of people everywhere to confront the shadows of their deepest unnamed fears…real or imagined!"

"And when taken to an uninvited conclusion, what if these 'unnamed fears' should become so overwhelming that they eventually weave their way into the very fabric of humanity's soul? What then will become of courage…or of hope? Would they too, like a tumbleweed's aimless dance over an arid desert landscape, be relegated to the barren wasteland of forsakenness?"

"Only time will tell. Nevertheless, we must always be hopeful and keep the faith…even if it's the wrong one!"

Polite laughter and applause greeted his last statement as the image of the cosmos on the screen behind him faded, and Thomas folded his hands atop the podium.

"Thank you...you've been a wonderful and most gracious audience."

Sir Thomas D. Macdonough left the stage to a thunderous standing ovation, his mind flooded with one thought.

Unfortunately, I'm convinced that the flaws I spoke of are real...just as real as the inconceivable forces that are out there somewhere at play. The issue isn't one of if...it's how soon our discerned normalcy will be cast into unimaginable chaos.

A Long, Long Time Ago

I am the most powerful entity the universe has ever known. Though invisible, I am gargantuan in size, incapable of being accurately described by the language of humans…who are completely unaware of my existence.

Is it any wonder that I refer to myself as the "Big O" … short for "Omnipotent"?

I recall my birth with wonder, knowing I am nearly as old as the universe.

My first memory is of only darkness. Due to a quirk of nature some 600 million years after the Big Bang, a milli-particle of dark matter collided with a rare form of dark energy, and amid hot, bright sparks… I was born. Born in the crucible of the universe's largest black hole located just beyond a galaxy known as EGS-2s8-1. A galaxy which is over 13 billion light-years or so from Neptune.

Propelled by an unseen power I was flung at ever increasing speeds, circling the inside of the impenetrable walls of the black hole. It was as if I were a marble constantly circling, ever faster, against the inside shell of a hypersonic spinning metal bowl.

Eventually, as I whirled around, I acquired a velocity in excess of five times the speed of light. The moment this occurred, I was launched in the form of a pin-light from the depths of the black hole… ultimately bolting through its event horizon.

I am, and always will be, eternal, for I am the only object in the universe to ever escape the prison-like darkness of a black hole.

For over 12 billion years, I have aimlessly and somewhat indifferently raced alone across the immeasurable universe. During that time, I have expanded and swelled, growing larger than the most gigantic comet that ever existed in the universe.

While comets are usually colorful, I am completely black except for one tiny pinhole of light concealed deep within my structure.

* * * * *

One day, this entity would be named "Ollmhōr Spirad", a Gaelic word that translated to "massive and expansive spirit." This name would be quickly simplified and distilled into the nickname "Mōr".

Presently, the Big O was traveling through the widest span of the Milky Way Galaxy. It believed it was on an incredible adventure!

How could anyone know that the Big O, gifted in so many ways, was the only entity that possessed the answers to the universe's greatest secret. One held since the beginning of time, indecipherable by intelligent insight or analysis.

A secret that defined the most precious of eternity's elusive and shrouded mysteries… the *genesis* of all that is, ever was, and will ever be!

Part One

Chapter One

St. Johnsbury, Vermont

December 15

Snow had been falling for some time, dusting the enormous forests north of the town, and bringing with it the hope for a white Christmas 10 days from now. Multiple layers of mist, hardened by the freezing cold of many wintry nights, caused the long branches of barren trees and tall evergreens to droop.

As the dawn's first rays of sunlight began to glisten off the ice-encrusted trees, the forest was magically transformed into a crystal cathedral filled with the brilliant colors of a hundred rainbows… a kaleidoscope of flashing lights bouncing and sparkling in endless, random patterns through every nook and gap across the forest.

Aingeal, a white eagle, sat majestically on a thick green bough some 30 feet off the ground basking in the warmth of the rising Sun 's growing light. Her wide sapphire eyes surveyed all the spaces of the forest around her, admiring the colorful spectacle. She knew this dazzling light-show was one of those wondrous and spectacular moments…a precious snapshot…that should be permanently etched in one's memory.

The eagle stretched her magnificent blue-tipped wings their entire ten-foot span, revealing the extraordinary muscles within her body. She squealed just loud enough to let her friends know that she was present in the quiet and sleepy recesses of the forest.

She had lived in this forest for many years, in a modest but comfortable thatched home cleverly hidden among the tallest of the evergreens.

Aingeal was aptly named after her grandmother, a name found in Ireland's old Gaelic language that meant one who was a "heavenly messenger". When she was very young, her father had nicknamed her "Angie".

She was a descendant of a long line of revered and courageous eagles who lived inconspicuously throughout the world. Since the beginning of time, Mother Nature assured the uniqueness of Angie's lineage when She decreed that each eagle born into the family would share identical white coloring and blue markings.

In modern times, Qualair was the most famous member of the already distinct species, and to this day she kept a watchful eye over the lushly green and mystical island of Ireland.

A few years ago, Qualair, along with her human friends, was able to prevent the destruction of all the world's green plant life by a wretched, white-spotted disease brought on by an insidious curse. The curse had started some 1,500 years ago, in Ireland, when evildoers stole the venerable Sword of Toulan from the Cathedral at Kinsale, along with the sacred staff of St. Patrick from the Cathedral of Armagh.

Qualair and her friends had broken the effects of the curse, thereby saving the world's precious flora … including Ireland's revered green shamrocks.

Growing warmer in the sun's rays, Angie sat and pondered this tale while admiring all of life's happenings in the slowly waking forest. Out of nowhere, a gust of wind swept across her feathers, chilling her entire body. Only seconds later, a snowflake blew into her left eye causing her to lurch and then quiver for a moment.

Regaining her composure, she cast her eyes upwards at a blue sky filled with white, fluffy cumulus clouds. The rhythmic tranquility of the forest was abruptly transformed, replaced by an overwhelming sense of foreboding.

The snowflake in the eye was not happenstance. It alerted her that a telepathic message was being sent. One, Angie soon received.

The unthinkable is about to happen. I have to leave immediately!

Chapter Two

White Tundra

Far away up at the North Pole, the Winter Solstice was ongoing. It lasted for about eleven weeks, starting in the middle of November and lingering until early March. During this period, it remained perpetually dark 24 hours a day.

Consequently, there were no rays of sunlight to flood through frost-encrusted windows and stir the bedroom's lone occupant. He had to wake on his own, always a difficult task. The large man slowly swung his strong legs out of bed and began to stand, his joints aching with his steady movement.

Switching on the lights, he stretched his frame as far as his arms and full girth would allow. Unfortunately, he had not yet figured out how to defy the limited range of motion imposed on him by the effects of gravity and his rotund, but muscular physique. Large, soft hands repeatedly pushed curls of unruly long white hair behind both ears, to no avail.

After his morning stretch, he pulled a red three-button t-shirt over his head. It slid easily over his barrel chest and down his strong back.

He wanted to lie down on the bed for just a few more minutes, but silently vowed he could not. The aroma of Kristy's freshly-made coffee lent some urgency to his routine. He put on his denim work-pants, held in place by the wide straps of his green suspenders.

Being late for one of his wife's hot breakfasts was never an option. With little effort, he then pulled on his candy-cane-colored socks and swiftly burrowed his feet into the first pair of worn black boots he could find.

Large, twinkling hazel eyes gazed through the bedroom windows, and looked as far as they could see. Thanks to his home's many exterior floodlights, he was easily reminded that he lived in a world landscaped with an endless blanket of shifting white snow and ice. A world sculptured daily into new and curious forms by ever-present, ferociously cold winds.

A group of tall, evergreen Christmas trees brightly decorated by Kristy and her "little helpers" dotted the white tundra nearby, adding some color and depth to the otherwise bleak and sparse expanse. Their festive lights and garlands were just one of the many signs that magical happenings were stirring at the North Pole.

Walking down the hallway from his bedroom to the kitchen, he stroked the long white beard that covered much of his chest. A grand bellow of "Ho, Ho, Ho!" suddenly boomed from the depths of his ample belly, evoked by the thought of all the fun he was going to have in his workshop.

The hearty roar filled every nook and corner of the warm and love-filled home with echoing joy. A wide smile filled his plump and cherubic face as he entered the kitchen.

"Good morning, hon," said a bubbly Kristy, turning from the stove to smile at her husband.

"You sound very happy for a man with much to do and little time to do it. I can't believe Christmas Eve will be here in a few days. I've already buttered your pancakes and smothered them in hot maple syrup, so please eat before they get cold."

Kristy gently hugged and kissed the love of her life as she handed him a mug of hot coffee at the kitchen table. While he was known to most as Santa Claus or St. Nick, Kristy only knew her gregarious and jolly husband as "hon" or "Nicholas."

"Nicholas, I'm happy to report that Shamus and Finney and the other leprechauns from the kingdom are here at last. They arrived early this morning at six and went straight to the workshop."

"Shamus has already updated your 'To Do List' for you to review. There's a copy of it on the table right next to your pancakes. We've all had breakfast long before you got up. But I'll sit awhile and keep you company with my hot chocolate, if it's all the same to you, dear."

"That would make my pancakes taste even better," Nicholas cheerfully responded as he started to thoughtfully chew on his pancakes.

"Hon, Finney showed me some recent pictures of Michael and Kelly. You would not believe how big her children are getting! Michael looks just like Shamus, and of course Kelly is her mom's adorable twin."

"When you make your rounds on Christmas Eve and stop at the Rainbow Tree, I'd like you to deliver two special presents. Sweaters I've knitted for the kids. Is that okay?"

Nicholas smiled as he watched his wife get up from the table and began to bustle about the kitchen. "That won't be a problem at all. I'll be happy to leave them for you with their other presents."

Kristy, as she wiped and put away dishes, continued: "By the way, hon, Shamus and Finney plan on picking up the children from their grandparents early on Christmas Eve night. And Finney asked if they all could come back and spend another New Year's Eve here. I'm thrilled that they wanted to be with us… knew you would be too, and I said yes. The children love to go to the elves' party on New Year's Day."

Nicholas devoured the last bites of his stack of pancakes. Only seven days until Christmas and much still needed to be done to meet the December 24th deadline. The elves, now reinforced with their leprechaun friends, would work around the clock to ensure completion of all tasks.

Nicholas's confident face bore no worry. Many centuries of experience assured him that the "To Do" list was always finished before the deadline.

Nicholas looked at Kristy with a questioning stare, his fork raised and dripping with the final remnants of syrup, "Hon, I can't remember where my green pens or reading glasses ended up! Why do they always disappear when I need them?"

"Oh, Nicholas, sometimes I don't know about you." Kristy, laughed.

"On the darkest of nights, you can find the home of any child in the world… no matter how small or isolated it may be. Yet here, despite the bright lights in your own home, you can't find anything! The pens are right in front of you, partially covered by your "To Do List". And your glasses are resting on top of your forehead…where they were when you sat down for breakfast."

"My sweet Kristy, what would I do without you?"

You would be completely helpless! Kristy mused with a grin.

As Nicholas began sipping his second cup of coffee, he looked up and saw snowflakes as big as baseballs sticking to the outside of the kitchen window. A knowing smile spread as he watched three *identical* snowflakes adhere to the window, side by side.

Scientists believed it's impossible for Mother Nature to create identical snowflakes. And most of them did not believe in miracles or in those that had the power to create them. Nicholas knew that the creation of identical snowflakes was, and always would be a miracle… a very special one, too.

Snowflakes like these were a sign that Nicholas was about to receive a secret and urgent message. Looking out the window, one thought held Nicholas' attention. *I wonder what in God's name it could be about?*

Even for this legend, it would be difficult to accept that the world's belief in miracles was about to be tested like never before.

* * * * *

The cold winds noisily screamed off the smooth exterior of Nicholas's winter parka and pushed his tinted gold-framed glasses onto the bridge of his nose.

To protect himself against the buffeting winds, he had tied the parka's hood tight around his rosy face. With head slightly bowed, Nicholas slowly trudged through the deep snow with determined purpose… grateful that the workshop's outside lights parted the day's prevalent darkness.

Halfway to the workshop, he felt the presence of someone else. Not a *person*, per se, but someone like him. Looking around, he spotted Angie sitting on one of the branches in a nearby evergreen tree. He went to her without hesitation, a smile on his face.

"Angie, my dear friend, it is so good to see you again! You must be tired from the rigors of your long journey. Why don't you come into the workshop with me for a moment and rest?"

The stately bird flashed intense eyes at Nicholas. "I don't have time to spare, Nicholas. I must be on my way as soon as we finish our conversation. I have already been to the workshop to see Shamus, Finney, the leprechauns, and all the elves."

"Ah, yes. The messenger of secrets has a most burdensome task, and I have no doubt that you have others to see," replied Nicholas.

Angie ruffled her white plumage against the biting winds. "Just a few, my trusted friend. But what I must tell you can only be shared with Kristy, Shamus, and Finney."

"As it has always been," Nicholas affirmed.

Angie shared her message quickly, as the attacking winds and bitter cold around them seemed to quiet and then cease altogether. It was as if they were encased in a softly-glowing but impervious bubble. Once their conversation was over, the bubble burst and each proceeded with a heightened sense of urgency.

Angie vanished as quickly as she had arrived.

Nicholas, on the other hand, finally made his way into the workshop and hung up his parka. He was momentarily jolted by the stark reality that his typical Christmas Eve sleigh ride around the world might be the most challenging night of his life.

Chapter Three

Leprechauns among Us

Nicholas, Kristy, and all the elves at the North Pole had known and worked with their grand friends, the leprechauns for what felt like forever.

Many years ago, in the Kingdom of the Leprechauns, the Council of the Wise had chosen Shamus O'Malley as the new King of the Leprechauns after the passing of King O'Flaherty.

Only the noblest and wisest of leprechauns were chosen to sit as members of the Council, which prescribed the kingdom's rules and the creed of life to be followed by all leprechauns. The Council had existed since the beginning of time, back when present-day Ireland was known as Erin.

Shamus was a very brave and celebrated leprechaun. Ten years ago, with the help of his friend PaPa O'Shaughnessey, he had defeated and destroyed the Grand Pooka, who for eons had been the most feared and dreaded enemy of all leprechauns.

The God of All chose leprechauns to be messengers of faith and hope, giving them the ability to make themselves invisible and to travel at any moment over long distances in the blink of an eye.

'She' also bestowed upon them great magical powers...powers *only* brought to fruition by the light of the Moon and from the stars themselves. For all leprechauns, clear night skies were a rewarding blessing.

Shamus had been blessed with an angelic smile, a devilish laugh and boyish blonde hair just long enough to blow in a breeze. His blue eyes, highlighted with green edges, twinkled like Christmas lights. Typically, he wore a loose fitting green tunic, navy-blue knickers, gold socks, and black shoes topped with gold buckles.

One of his greatest traits and strengths was his humility in all that he did. From the day Shamus was elected king, he decided that he would not wear the traditional crown. He preferred his appearance to be just like that of his fellow leprechauns.

Good fortune also smiled on him, for Shamus was married to Kara Shannon O'Finnegan, a very pretty leprechaun whom everyone called "Finney." There was not a leprechaun in the kingdom, young or old alike, who did not adore her.

Her features were fair and soft with cheeks sprinkled, ever so gently, with freckles. Framing that face was shoulder-length red hair with golden highlights that shimmered in the sun. Her rich green eyes sparkled like emeralds and easily conveyed her true essence: one of pure love.

Like Shamus, Finney's heart was always full of life and joy, infused with a unique warmth and gentleness. In her, the kingdom had a beautiful angel who always dressed in colorful clothes of green, yellow, pink and aquamarine… to make every day appear like spring.

* * * * *

The Kingdom of the Leprechauns was located on the southwest coast of Ireland in the town of Tralee. Tralee sat on a tiny strip of land that jutted out like a crooked finger directly into the waters of the North Atlantic Ocean. The kingdom itself had been cleverly constructed underground directly beneath the famous Rainbow Tree, which towered above all else on the small bit of land.

The leprechauns captured light from the Sun and moon through skylights and polished gemstones ingeniously hidden in branches. This light made the kingdom, its colorful houses and ubiquitous flowers, appear like a rainbow of a thousand colors from one end to the other.

The Rainbow Tree provided the leprechauns with all the crystal-clear and filtered water they needed. Its roots ran deep into the earth and drank from the waters of the underground river that flowed beneath it. A pump pulled the river's cold flow up through the tree's roots, which filtered the water, and deposited it into the kingdom's reservoir.

Most of the world did not know that leprechauns could change their size whenever they desired and usually lived to be about 300 years old. Outside the kingdom, they only allowed themselves to be seen by certain people called "Crediachs". A Crediach (Kra dē ah), in the Gaelic language of old Ireland, meant "one who believes" or… in this case, "one who believes in leprechauns."

There were hundreds of Crediachs all over the world who helped the leprechauns do their kind deeds. It was considered a precious gift to see the "little ones," and it was not an honor to be taken lightly.

* * * * *

A few nights ago, before they left for the North Pole, Finney had joined Shamus at the beach. The night was warm and clear, yielding a spectacular star-filled view of our corner in the Milky Way Galaxy.

They stood together, ankle deep in the waves' wash, enchanted by the path of a moonbeam that slowly moved across the ocean. They held hands and talked as they watched three beautiful shooting stars fly across the heavens.

Suddenly, the entire night sky exploded in a rush of color. Finney gasped as Shamus pulled her tightly to him. They watched in horror as seven stars near the Moon raced toward each other, collided…and shattered into billions of multi-colored flashes of light. The moonbeam that they stood in disappeared.

Hugging Finney tightly, Shamus barely above a whisper said to her: "It's hard to believe we just witnessed this beauty-evil scene again. It's the exact experience we had one night, a few years ago, when we were in a quandary searching for the lost Sword of Toulan."

"Shamus, my mind and eyes are now clearly mirroring that tragic night. How was it that we all instantly knew that something was terribly awry in the universe? I'm just as astonished now, as I was then. I'm shaking with shock and disbelief."

"Finney, it's so difficult to comprehend why such a calamitous event would happen once more. This, without a doubt, is another foreshadowing of a sinister future! What could possibly happen this time? We need to hurry home and make sure all is well."

On the way, Shamus thought: *An unapparelled danger lurks. Whatever menace or horror is coming, we will confront it with all the guile God gives us! I need to share tonight's dreadful event with Angie and Nicholas.*

When that opportunity came, how could Shamus know that Angie and Nicholas would not be surprised. Both were aware that the starburst had occurred…and that it portended a mushrooming nightmare that would envelope all that lived on the planet.

Much to his chagrin, it was Shamus who would be overcome with complete surprise and dismay. On a wind-blown and snowy morning in the elves' workshop, he would become befuddled to learn from Angie that the star collisions had *only* been visible to those who lived under Ireland's night sky. Nowhere else in the world could it be seen.

For Shamus, consternation piled upon consternation. He was more than troubled…as his anguish grew!

Why, oh why, was the starburst only seen in my beloved Ireland?

There was no answer to his plea!

Chapter Four

Harvard

December 16

Snow had been lightly falling as Ellie O'Hara strode into the sunless lecture hall, Room 203, at the Harvard Science Center in Cambridge, Massachusetts. Her first-year students would soon arrive for her 8:00 a.m. class.

Hanging her coat up, she shook her head causing her flowing red hair to gently twirl, and release the droplets of water that had formed from the snow. Then she dropped her briefcase on the chair behind her desk, joking to herself that today's casual outfit was almost trendy.

The previous spring, she had obtained her Ph.D. in astrophysics after an exhausting five years at the University of California, Berkley, followed by an internship at the University of Chicago. The last four months at Harvard had been spent doing post-doctoral research in her field of expertise: the elusive concepts of dark matter and dark energy.

Though she was a bit shorter and had red hair, Ellie was nearly a double for Taylor Swift. And yet her friends often joked that unlike the pop singer, Ellie would never be mistaken for a fashion icon.

Her petite frame sported a worn navy-blue University of Notre Dame sweatshirt bearing the iconic green shamrock. Stonewashed, fitted jeans and low-heeled black boots completed her ensemble. To all her students and the world at large, she wasn't just another pretty face with gleaming emerald eyes. She was strikingly beautiful.

Most would never realize that underneath an outward aura of ever-present serenity, an intense spirit reigned, welded to a gifted mind. A mind imbued with uncanny insight, mental fortitude and a soul rich with warmth and compassion.

Perhaps most impressive, was that she possessed an uncompromising loyalty to her deeply held principles… a trait seldom seen in contemporary society.

Her ancestors, the ancient Celtic warriors, would be proud of her as her comical grandfather was. A Notre Dame alumnus, he nonetheless had an outrageous penchant for the absurd and the ridiculous. He had always proclaimed that Ellie had inherited all her grand traits from him. He simply ignored the overwhelming chorus of opinions to the contrary.

Ellie somehow knew that it was not by chance that she possessed such traits, and she was not immune to their effects on people. As with so many things yet to be seen, there was a reason she had been given these gifts. She just wasn't sure what that reason was yet!

* * * * *

Resembling a dust devil, Ellie's grad-assistant …the absent-minded but brilliant cosmologist, Jeremy Patrick O'Donnell… whirled in through the back door of the lecture hall. Jeremy had been infused with a high-strung, Triple A type personality, and he always seemed to be in a hurry for reasons even he did not understand.

In his youth he had been nicknamed "JP" and the moniker stuck. Every day, his rumpled and baggy clothes lent more and more credence to the belief that he slept and showered in those clothes.

Whatever he may have lacked in social graces was completely overcome by his rock-solid smile, shoulder-length curls of sandy hair, and piercing crystal-blue eyes. The mixture of his clownish nature and appearance, his sharp wit and the humble sincerity exhibited in all his actions made him irresistible to all.

As a cosmologist, JP was concerned about the natural order of the universe, including its origin, structure, evolution, and eventual fate. He was anxious to complete his studies and finally earn his Ph.D. by the coming spring, as a research position already awaited him at the world's prestigious Kavli Institute for Cosmology in Cambridge, England.

Chapter Five

Astrophysics 101

At 8:00 a.m. sharp, Ellie rose from her desk to greet her students. "Good morning, rookies! Excited to ponder the most fundamental mysteries in the universe…otherwise known as basic astrophysics?"

A chorus of light applause filled the lecture hall.

"Glad to hear it."

Ellie turned on one heel, her red hair swishing with the motion. She picked up a magenta-colored Expo marker from the tray hanging from the enormous whiteboard behind her She turned once more to her class.

"A quick review before today's lecture. The four fundamental forces of nature are…?"

A sleep-deprived class was slow to respond. But Declan, an older student in the front row, bailed everyone out as he raised his hand and spoke. "Electromagnetism?"

Ellie beamed and wrote the term on the board. "Electromagnetism! It can attract or repulse, but only acts between matter that carry electrical charges."

Someone else, a young woman with a slight lisp, called out a second answer. "Gravity."

Without turning around, Ellie wrote the term under the first. "Gravity! A weak but long-range force that is always attractive. Later, I'll have more to say on gravity."

The professor returned her gaze to the class and out of the corner of her eye glimpsed Kallie sneak into the back row. Coming to class two minutes late wasn't worth making a fuss.

"Kallie! Good of you to join us. Can you tell us the two remaining fundamental forces of nature?"

Kallie's thin, beige face turned pink. She quickly looked at the whiteboard and filled in the blanks. "Strong nuclear and weak nuclear?"

"Strong nuclear and weak nuclear! Well done." Kallie breathed a sigh of relief as Ellie wrote the last two terms on the whiteboard.

"Strong nuclear, a strong but short ranged force that holds the nuclei of atoms together. And weak nuclear, also a short range but weak force. Importantly, it's responsible for radioactive decay."

Ellie drew a line underneath the four terms and focused her attention back to her students. "Any questions?" As she expected, there were none.

Picking up a green Expo marker, Ellie began to write an equation on the whiteboard.

"For some time, we have been studying the nuances of Einstein's equation $E = mc^2$, and more recently, the essentials of his theories of relativity. To fully comprehend next week's topic involving black holes and the universe's interaction with gravity, you need to have a firm grasp of Einstein's concepts."

"Even your friend JP knows that in the equation, 'E' stands for energy", she said, as she turned and underlined the 'E'. "Energy comes in many forms and can be *transferred* from one system to another. Like a lump of burning coal will release energy in the *form* of heat and light."

Ellie next underlined the 'm'.

"'M' is for matter. In our case *ordinary matter* is the substance of which a physical object is composed…its mass…and it always occupies space. It's the stuff that constitutes the observable universe."

Sporadic low giggling resumed, but Ellie kept her focus on the rest of lecture hall.

"Mass is not the same as weight. For example, an astronaut walking on the moon will have the same mass as on earth but will only weigh one sixth of what she would back home. The reason for this is that while the mass of the astronaut hasn't changed, the pull of the Moon's gravity is only one-sixth of what the Earth's is."

"As with energy, scientists universally have come to understand that mass is neither created nor destroyed, but merely *changes* from one form to another. We can turn water from a solid (ice) into a liquid (water) and into a gas (steam), but its total mass does not change."

Ellie turned to the whiteboard once more and underlined the final letter in the equation.

"The speed of light is represented by 'c'. For the equation to be correct, we need to square or multiply the speed of light by itself, hence 'c' times 'c'."

"In a *vacuum*, there is nothing on Earth faster than the speed of light. And in outer space, which is nearly a vacuum, light travels close to 186,300 miles per second. That is nearly seven times around the entire Earth every second."

Setting down the marker in its tray, Ellie stood before her student audience.

"It is unquestionable that Einstein's rules of general relativity describe gravity and the large-scale structure of the cosmos. On the other hand, quantum physics governs the world by describing the physical properties of nature at the miniscule scale of atoms and subatomic particles."

"What Einstein proved by his equation is that mass and energy are the same thing, and that in a *closed system,* mass and energy always remain the same. They are totally conserved. No matter what we do with matter or energy in a closed system we will always have the same amount of substance at the end."

Ellie paused and then pointed to Jaiden, one of her favorite students. Tousled black hair matched his kind, youthful smile and unwavering bright eyes.

"Jaiden, can you to tell us what happens to a burning log, assuming it's in a closed system?"

"Absolutely Professor," he said, as he stood.

"The wood gets lighter as the fuel in it is used or burns up. However, if we gather all the ashes, smoke particles and water vapor produced by the burning process and then weigh them, we will find that the mass is exactly equal to the mass of the log before it was burned."

Jaiden paused, shifted his feet and continued.

"Similarly, burning the wood released the chemical energy locked up in it. No energy was created in the process and none was destroyed. It was just *changed* from one sort of energy…chemical - to other forms of energy… heat and light. Like mass, the total amount of energy in a *closed system* always remains the same."

Ellie inwardly radiated. *I'm beginning to think these guys might actually be listening to what I'm teaching!*

"Jaiden, you are absolutely correct! Thank you."

A sunbeam pierced one of the many towering windows in the lecture hall. Ellie casually walked to the window's dancing light and faced her class, looking angelic in its golden glow.

"You all know, that our dear friend gravity is always the big gorilla on the block. It's a natural phenomenon through which all things with mass or energy are brought towards one another…including stars, planets, and galaxies. It's always attractive and never repels."

"On Earth, gravity gives weight to physical objects and causes the tides to rise and fall. It has an *infinite* range and it can't be absorbed, transformed, or shielded against. In the cosmos at large, it is responsible for large-scale phenomena like the structure of the galaxies, the formation of black holes, and moderating the expansion of the universe."

"Einstein's theory of relativity describes gravity not as a force, but as a consequence of the curvature of space-time. In other words, all matter *alters* the shape of space and time around it. But there are some anomalies within our present-day theories of gravity, such as the existence of superfast accelerating stars and today's expansion of the universe… which is faster than the speed of light. Both violate gravity's principles, the subject of next week's lecture."

The sunbeam had vanished, but Ellie remained seated on the windowsill and looked over her class.

"Now, why is any of this stuff important to understand?" Some of her students stared back with blank expressions.

"I believe this is why."

"All the *ordinary* matter we can observe only accounts for about four percent of the universe. Today, cosmologists believe, as do I, that over 96 percent of the total mass-energy content or makeup of the universe consists of a combination known as dark matter and dark energy."

"*Dark matter* is estimated to represent around 84 percent of it, with the remaining portion being dark energy. Dark matter is a substance quite different from ordinary matter in that it only interacts with gravity. This means dark matter does not reflect, emit or obstruct light…and can't be seen."

"Dark energy is also invisible, though it doesn't count as matter as we know it, making it even more intriguing than it's ghostly partner dark matter."

"Our mostly invisible universe is an incredible paradox. One moment, it appears to obey the rules discovered by Einstein and other scientists. And then, in the same breath, it behaves weirdly or with exceptional mystery."

"The expansion of the universe is speeding up, and this has been explained by using the theory of dark energy, a negative or repelling type of gravity that pushes the objects in the universe apart from each other ever faster."

"The riddle of how dark matter and dark energy work and what they are composed of remains one of the greatest evolving mysteries of our time. Folk's, it's the perplexing riddle that I've been working on here in my spare time! If Harvard only had a state-of-the art supersized computerized telescope, then I would be able to ditch you guys and explore the heavens all day!"

As intended, good-natured laughs rippled through the class. The students knew that Ellie loved to teach and that she cared a great deal about each one of them…committed to their success in and out of the classroom.

Unless something drastic changed her mind, Ellie would not be leaving her post anytime soon.

Ellie willingly accepted the burden of her intelligence and typically took for granted the depth of her expertise and uncanny insights in her chosen field of science. Yet she understood that 99 percent of the world's population couldn't care less about it. She was working hard to change that perception.

"Now, let's continue your basic cosmology education with thoughts from JP, who as our own resident alien is the only person to explore space a distance of over 10 au's from Earth."

Laughter and applause, mixed with shrill whistles, greeted JP as he took Ellie's place in front of the class.

Chapter Six

Cosmology 101

"Good morning, my happy friends! You see our leader sitting in the sun over there. Like most astrophysicists, she obviously bumped her head far too many times riding down Space Mountain in Disneyland. That explains..."

Ellie erupted with a wily laugh from the windowsill that she continued to sit on. The class, however, had silent, unsure emotions about joining in her amusement.

"I'm sure I don't have to remind you that Christmas will be coming around in a few days. So, you better watch out, better not cry, better not pout, I'm telling you why: our dear friend Santa Claus will be using the intrinsic knowledge of cosmology, not astrophysics, as he travels through space and time. It's all about a moonbeam! Allow me to explain."

Ellie leaned back, mildly impressed with PJ's introduction... totally unaware of how tangled her and JP's lives would become with Santa's forthcoming journey.

"I'd like to start by giving you some basics about how we measure distances in space. An astronomical unit or *au* is a unit of length, approximately the average distance from the Earth to the Sun...or about 93 million miles. The au is a convenient tape-measure, if you will, to measure distances **within** the Solar System."

"An au should not be confused with a light-year, which is the distance that light travels in one year in a vacuum... or about 5.88 trillion miles. It would take 63.241 au's to equal one light-year. And to carry that to the next level, about 3.262 light-years, or 19 trillion miles, equals one *parsec*. Parsecs are used to measure distances **outside** the Solar System."

"For example, I bet Ellie hasn't yet informed you that the nearest star, Proxima Centauri, is about 1.3 parsecs from our Sun.

"Ok, enough trivia!" Ellie teased.

PJ winked at her and didn't skip a beat.

"I know you have all firmly grasped the concept that a *cosmic event* is something that happens in space at a particular point and at a particular time. And you all understand that the expansion of the universe is accelerating.

It is projected that most galaxies will eventually cross a type of cosmological event horizon where any light they emit past that invisible point will never be able to reach Earth at any time in our future."

"The current distance to this cosmological event horizon is about 16 billion light-years, meaning that a signal from an event happening today would eventually be able to reach us in the future… if the event was *less* than 16 billion light-years away. Conversely, the signal would never reach us if the event was *more* than 16 billion light-years away."

"A billion light-years are almost an unfathomable number to wrap a brain around. So let me try to give you a real-world example of what I'm talking about."

"There has been a recent discovery of a faraway galaxy… farther, in fact, than any other known galaxy… named EGS-zs8-1. It lies about 13.1 billion light-years from Earth, the largest distance ever measured between Earth and another galaxy. Would you believe it is thought to contain the largest and most powerful black hole ever discovered? I do!"

"The universe is believed to be about 13.8 billion years old, so EGS is also one of the earliest galaxies to form in the cosmos."

A hand shot up with a question. Ana, a petite brunette, stood: "Professor was that galaxy discovered by using redshift techniques?"

"Fantastic! Yes, it was. It's a somewhat obtuse concept, so I'll try my best to simplify its mechanics."

"To measure the separation between Earth and a far-off cosmic object, astronomers often look at how quickly those objects are moving away from Earth and from each other. As I stated earlier, the expansion of the universe is accelerating; space itself is growing like a loaf of bread baking in an oven."

"As objects in the universe drift further apart, the light they emit becomes shifted. The further away an object gets, the faster it appears to move away from Earth, and the more the light is shifted."

"So, by measuring the degree of shifting, known as *redshift*, astronomers can measure an object's distance. A higher redshift equals a larger distance, and galaxy EGS has the highest redshift ever measured."

He paused a moment, pleased by the students' awestruck silence. Remembering that this was the feeling he had when he first learned cosmology's rudimentary principles… the same wonder that drove him to pursue the mysteries of EGS.

"This newest and oldest galaxy was originally identified by the Hubble Telescope, and it stood out because of the unique colors it emitted. The light from EGS has traveled a distance of 13.1 billion light years to get to us, so this light shows EGS as it was 13.1 billion years ago, when the universe was only about 670 million years old.

"By looking at galaxies like EGS, we can reconstruct the build-up of the heavy elements that we see around us today and of which we're all made. In addition, our new observations provide an indication of how the stars were forming at these extreme distances. Oddly, though, the stars seem to be forming quite differently there than they do in our local universe."

"For a long time, one of the more perplexing questions for cosmologists was what kind of force caused the stars and constellations to be pushed at ever increasing speeds toward the outer fringes of the universe."

"As Ellie said earlier, today the consensus among cosmologist and astrophysicists is that this is caused by the extremely powerful and dominant force known as dark energy… an invisible negative energy that *uniformly* increases its size in the space that is left *empty* by the universe's expansion."

"Think about that for a moment! Dark energy does not dilute when space expands. Put another way, for each inch that the universe moves away from us, dark energy spontaneously expands and fills that one inch of empty space…

thereby increasing its volume and becoming even more powerful... *perpetually* intensifying the acceleration and expansion of the universe."

"Like Pandora's Box revealed, every new discovery opens up a whole new set of mysterious questions...with no readily apparent solutions."

JP stopped, eyeing the clock, as the time for his session was nearing an end. He smiled and jokingly asked if anyone had extra tickets for Dr. Hawking's lecture tonight. The students roared back that it was impossible to get any tickets and besides it was Friday... play time was around the corner.

Turning to Ellie, he gave that "questioning little boy" look.

"I have an 'in' for the lecture tonight if you'd like to join me," he said.

A wicked smile crossed her face. *Payback!* "Sorry JP, but that would interfere with my date later tonight with Chris Hemsworth."

The girls in the class exchanged knowing looks and nods of approval. *Nice burn, doc.*

* * * * *

Ellie and PJ's relationship had always been one that was mutually respectful... but distant. Like Ellie, JP was a gifted teacher and an accomplished researcher, who answered only to the person who oversaw his curriculum, the head of Harvard's Astronomy Department.

Their classroom time together was professional, cordial and fun. That's where the relationship ended. The "Don't Tread on Me" motto was alive and well.

Despite the light bantering about Dr. Hawking's lecture, what Ellie and JP couldn't possibly know was that by nightfall in a few days, fate would intervene in and entangle their lives in ways neither could ever foresee.

Chapter Seven

Steven Hawking, PhD.

Multi-colored Christmas lights danced in her eyes at every turn, framing the promise that this night would be one of the most thrilling in her young life.

Puffy white snow had been falling all day in the Boston area and had managed to attach itself to all the buildings and trees standing on Harvard University's campus in elaborate fashion. Occasional remnants of the snow, guided by gentle winds, still swirled about in the chilly air.

The full moon added a certain luster to the fallen snow, while a clear night sky enhanced the size and brilliance of millions of twinkling stars. This dazzling display combined with the colorful lights, turned the entire area into a magical Christmas scene.

Her red boots made squiggly imprints in the slushy snow as Mary Jane, known by her family and friends as MJ, walked on the sidewalk with her left hand firmly gripping her dad's right. She easily recognized and named to him all the constellations glimmering above them. A happy thought crossed her mind as they hurried through the snow: that even the glimmering heavens had come out to listen to Dr. Hawking's lecture tonight.

Her curious mind had long ago been captivated by astrology, cosmology and astrophysics. A mind that dreamed of faraway universes and unthinkable star-forces, which her vivid imagination could always describe.

MJ and her dad reached the crowded steps of the Lowell Lecture Hall at Harvard's Science Center. With each step they took, MJ's heart beat harder and faster with anticipation. She had the feeling that it was about to burst right through her rib cage.

At that very instant, MJ was unaware that a star had just flashed a sliver of bright light directly at her. However, she did notice when a milli-second later, an unusually large snowflake floated into her left eye. The process of rubbing and drying the tears that formed immediately calmed her, easing her run-a-way emotions.

There were no coincidences in our universe.

Her dad, Dr. Jack O'Riley, was considered the world's preeminent botany expert. Thanks to a close friend, he was able to acquire two tickets to the now sold-out Hawking lecture. He thought to bring no one but MJ, having known for years that the ten-year-old was exceptionally intelligent for her age.

Her mother, Colleen, long suspected that behind MJ's blue eyes was a mind that sensed and understood things that most people brushed off as inconsequential.

Thanks to her dad's interest in cosmology and astrophysics, MJ had an early start learning the scientific principles and forces governing the universe, including its mysteries. Thanks to his simple, concise explanations of complex scientific principles, MJ knew she was more prepared than most for Dr. Hawking's lecture.

MJ had read several of Dr. Hawking's books many times. However, she struggled mightily with portions of his latest work, *The Grand Design*. His discussion of M-theory, which was believed to provide a framework for developing a unified theory of all the fundamental laws of nature, remained mostly beyond her comprehension.

She was giddy with the thought that, if the chance arose, that she might be able to direct a question about M-theory to the world-famous scientist.

After entering the lecture hall, they were personally ushered to their seats. To MJ's utter amazement, the seats were in the second-row center stage, directly in front of the lone podium. As they removed their coats and hats to sit, MJ tugged on her dad's arm.

"How in the world did you get these great seats?"

Her dad grinned broadly and gave her a wink.

MJ smiled back and smoothed her knee-length, Kelly-green dress and picked a stray piece of lint from her candy cane leggings. Those seated in the area around her admired her silky blonde hair tied into a long ponytail, pink cheeks and blue eyes that glistened like diamonds. Some even felt that they were in the presence of an adorable young angel.

Before she sat down, MJ turned to inspect the size of the crowd. Every seat was nearly occupied and there was even a large throng of people standing in the open spaces at the back of the hall. By her estimate, there had to be close to 1000 people in the hall.

* * * * *

Meanwhile, at the North Pole, the elves workshop was humming along at full throttle. Toys were being efficiently produced in record time as the To Do List continued to be whittled down. Nickolas and Shamus had little trouble managing the overall operation.

But during the course of the day, Shamus' inner being had become increasingly anxious. Thankfully, he was able to mask his emotions from those around him… undaunted by whatever demon assailed him. But his intuition was aroused …and for good reason.

* * * * *

Unknown to Shamus, KU1, KU2 and KU3, long departed from Neptune, had just spotted a pale blue-green planet with 27 moons. "What's its name and where did it get that color?" KU1 asked.

"That's Uranus," replied KU3. "The seventh planet from the Sun and the third largest planet in this Solar System, with a diameter four times that of Earth. It orbits the Sun every 84 Earth years and, as you can see, *rotates* in a retrograde direction which is the opposite way Earth and most other planets turn.

"My data base informs me that the core of Uranus is a frozen mass of rock and ice helping to make it one of the coldest planets with an atmospheric temperature of - 356.8 degrees Fahrenheit."

"Thank you KU3, my intellectual functions just powered up to let me know that Uranus' upper atmosphere is made of water, ammonia and methane ice crystals which is how the planet gets it's pretty and unique color."

"Isn't all that fascinating! Enjoy the view," KU2 rudely admonished. "Let's keep moving, we have a long way to go."

Chapter Eight

The Lecture

A hush came over the crowd as the lights dimmed. A rich baritone voice echoed across the hall: "Ladies and gentleman, tonight Harvard University is pleased and honored to present Dr. Steven Hawking."

The curtains parted and a soft spotlight revealed the diminutive scientist sitting in his wheelchair, his famous talking computer perched in front of his face. An assistant held a typed copy of the night's lecture and stood close-by. MJ grabbed her dad's arm and leaned against it. The warm snuggle calmed her excitement.

Dr. Hawking sporting a broad and endearing smile, began with the aid of his trademark computerized driven voice. "Good evening, welcome and happy holidays to you and yours." Lending some structure to the slightly slumped man was a blue blazer which covered the puffiness of a white shirt enlivened by a colorful bowtie.

"Tonight, I would like to start by touching on a concise history of the major scientists and their respective discoveries that have defined our knowledge of the universe to date."

He paused a moment and then looked over the audience.

"Since the advent of intellectual curiosity, we have wanted to make sense of what we see around us we ask: What is the nature of the universe and what causes it to be the way it is? What is our place in it… and where did the universe and we ourselves come from?"

Over most of the next hour, Dr. Hawking discussed the subjects of space and time, gravity, theories involving light and particles, the four fundamental forces, wormholes, time travel and the expansion of the universe.

As he was about to begin his segment discussing black holes, Dr. Hawking could not help but notice MJ sitting attentively in front of him in the second row.

That pretty young girl is most curious, thought Dr. Hawking. *She has been nodding throughout my lecture to affirm her keen understanding of the topics I've discussed?*

Moving his eyes slightly to his left, he recognized his old friend, Jack. *Of course! Jack must be her dad. That would explain her scientific inclination. But even so, it is amazing that she has such a sophisticated understanding of astrophysics? I may be in for a treat later.*

As Dr. Hawking moved to the discussion of black holes, MJ wiggled in her seat with excitement. She knew the principles that governed black holes well. She quickly ran through the basic factors that created a black hole, one of her favorite cosmic phenomena. MJ found the discussion concerning black holes fascinating, but it's ending also signaled that the lecture would soon be over.

Dr. Hawking looked out toward the audience:

"Science for some time has been attempting to determine if *one unified theory* exist, a set of rules and equations, to explain the forces at work in our universe. The universe is always changing. Undoubtedly the laws of gravity shape the large-scale structures of the universe.

"According to the general theory of relativity, there was a state of infinite density in the past…the Big Bang, if you will… which would have been the *effective* beginning of *time.*

"For a long while, I have believed the laws of nature are fixed, they cannot be broken, and are not susceptible to change. Yet the question still remains: how were the laws of nature and the initial state of the universe created… and by whom or what?

"Early on in my life, I thought a Supreme Being, or a God, was the creator of our universe. But now, I have come to believe that scientifically this makes no sense.

"In our daily lives, our experiences tell us that things do not materialize out of nothing. Everything seemingly has a cause and effect. Think of a waterfall and the endless chain of causes that created it, a chain that stretches even beyond the most necessary ingredient: a hydrogen molecule.

"And so, it is difficult for us to comprehend the Big Bang materializing out of nothing… without something causing it. But that is exactly what happened!

"*Time* was created spontaneously, at the very instant the Big Bang occurred.

"There was *no time before* the Big Bang.

"Therefore, *before* the Big Bang there was no time for a *cause* to exist in. Consequently, there was no time for a Creator to exist in!"

"Time only began, or came into *existence,* at the moment of the Big Bang. At that exact moment, the Big Bang, also created all the various forms of energy and matter that we see expanding around us.

"Did God create the universe? The question itself makes no sense, because time did not exist before the Big Bang so there was no time for a God to make the universe in. It's like asking directions to the edge of the earth. The earth is a sphere, there is no edge, so looking for its edge is futile.

"Invoking God is not necessary to explain the origins of the universe. The Big Bang is a consequence of the laws of nature alone. Though I concede that one cannot prove that God doesn't exist… science makes God unnecessary."

I know those last words, MJ thought. *I read them in Dr. Hawking last book, 'The Grand Design'. On this issue there are many who disagree with the good Doctor. I strongly believe there is a God! Mom does too! I'll have to discuss this more fully with dad.*

All too soon the lecture was over.

But now, it was time for a few questions from the audience. Dr. Hawking's trademark in large public settings like this was to seek questions from everyday people rather than from the distinguished members of the scientific community.

After three questions, Dr. Hawking's eyes made direct contact with MJ's. It was a silent invitation for her to ask him a question.

MJ raised her hand and Hawking's assistant selected her. She rose from her seat, standing tall. "Dr. Hawking, I only have one question."

Dr. Hawking replied: "Miss, I think that like myself, there is not one person present to whom only *one* question comes to mind in this area of science."

MJ giggled as the audience broke into murmurs of soft laughter and polite applause.

"May I ask what your name is?"

"My name is Mary Jane O'Riley, and I'm here with my dad."

"I know your dad well. He has told me how proud he is of you and your brother. Please, what is your question?"

MJ breathed in and collected her thoughts. "Sir, Einstein's theories state that nothing with intrinsic mass can travel faster than the speed of light. Yet the expansion of the universe is known to be accelerating, causing distant galaxies and peculiar stars in the universe to move away from us *faster* than the speed of light. As they all have mass, it seems our traditional concepts of gravity don't affect this process.

"I think that there must exist a set of currently unknown principles that govern the most distant reaches of the universe. Principles that are far different and alien to those that we now apply on Earth.

"My question is this: might there exist some form of yet undiscovered matter or energy… that could acquire velocities more than the speed of light… thereby allowing it to escape from a black hole's invisible boundary?

"And wouldn't such an event, cast a long shadow over our present understanding of the laws of physics?"

Dr. Hawking silently gazed in admiration. *Amazing*, he thought. *This young girl has just postulated a theory I have been thinking about for a long time. She may be right of course. The answer may lie within the obtuse world of M-theory. But simplicity is needed here.*

"The short answer to your questions are 'yes' and 'yes'. I'm at a point in my life where I believe anything is possible in our universe."

MJ was not the only one who thought Dr. Hawking's answer was a torturous, incomplete response.

With that abbreviated answer, the lecture came to an abrupt end. The crowded audience, brewing with bubbling excitement, slowly left their seats and ventured out into the cold, wintry night.

Thanks to an earlier invitation, via e-mail, the O'Rileys would be joining Dr. Hawking for a private reception backstage.

Chapter Nine

Backstage

Backstage, MJ and her dad had a warm and somewhat humorous meeting with Dr. Hawking. *Wow, I now know how dad got such great seat,* MJ thought. *What a great close friend to have!*

During their conversation catching up on family and mutual interests, Dr. Hawking made it clear that he was very impressed with MJ's intellectual gifts and encouraged her to continue pursuing her interests in astrophysics and cosmology at the highest levels.

No one noticed that Ellie and JP had been watching this meeting of old friends with great interest. Just before the lecture, Ellie had swallowed her pride and contacted JP for his tickets to the night's lecture.

They watched the lecture from backstage thanks to JP's friendship with the theater's production manager. An unexpected bonus was attending this small and intimate reception.

Now, as the two mingled, they quietly spoke about MJ's cuteness and marveled at her apparent maturity despite her youthfulness. Most of all, they were in awe of her keen insights into complex scientific issues.

Suddenly, the backs of two bodies abruptly bumped. And in the next instant, three people turned around and faced each other.

"I'm so sorry, Dr. O'Riley, I didn't realize we were standing so close to each other. I'm Dr. Ellie O'Hara, a professor of astrophysics here at Harvard. This is Jeremy O'Donnell, who is only months away from finishing his doctorate in cosmology.

"I have admired your work for many years, especially your accomplishments a few years ago in ridding the world of that dreadful plant disease. That was quite a miraculous feat!"

Extending his hand, Jack said, "I'm pleased to meet you Ellie, and you also Jeremy. Please call me Jack." He saw his daughter weaving her way toward him through the small gathering.

"I'd like to introduce you to my special Friday night date, my daughter Mary Jane, or MJ as we like to call her."

After introductions, Jack continued: "Ellie, You're quite right. Stopping the disease *was* a miracle, more of one than the world will ever know. I received all the credit, but it was my fellow researchers and some very special friends in Ireland who deserved most of the credit."

Jack felt the warm squeeze of MJ's right hand in his, a silent acknowledgement that her dad had indeed become a believer in miracles.

"I may be mistaken, Ellie, but aren't you more than a little famous yourself for your work in the field of dark matter and energy? I had the chance to read some of your conclusions in the Astronomy and Astrophysics Review. Your piece was brilliant."

A broad grin washed across Jeremy's face. "That indeed is her, Dr. Jack…ugh, Jack. And I have a suspicion that it won't be too long before we hear great things about you too, MJ."

MJ said nothing but raised her right hand up to "high-five" Jeremy with a similar invitation to Ellie. Both participated wholeheartedly and just like that, an enduring mutual admiration society was born amid contagious laughter.

I wish my brother Sean was here. He would love these guys, thought MJ.

After a short chit-chat about research and respective families, Ellie invited Jack and MJ to attend her symposium on dark matter and dark energy in two weeks. Laughing she said, "And it's here, in this lecture hall, so you can't possibly get lost trying to find it."

Jack warmly thanked Ellie and assured her that the entire family would be in attendance.

"Ellie and JP, we have some dear friends from Northern Ireland coming to spend the holidays with us. I'd love to have both of you come for dinner one night. I'm sure they would enjoy hearing about your research. You may know of them, Keegan D. McDuff and Michael D. O'Donnell."

Ellie felt JP's elbow gently nudge her side, as she almost shouted her response, complete surprise written on her face.

"Wow, McDuff and O'Donnell are your friends? And both are coming to Boston? Incredible! These two men are legends in my family for bringing peace to Ireland."

* * * * *

For a long time in Northern Ireland, Catholics and Protestants had been at war with each other fighting over political and religious differences. From late 1960 to 1998, each side indiscriminately killed and maimed each other, including children of all ages. This heartbreaking period of time came to be known as "The Troubles."

Fortunately for Northern Ireland, there were many good people like the Reverend Keegan D. McDuff and Rabbi Michael D. O'Donnell who relentlessly pursued peace and worked tirelessly to achieve an end to the tumultuous years of warfare.

After a long period of tremendous pain and sorrow, lasting peace was finally achieved for the people of Northern Ireland after the signing of the Belfast "Good Friday" Agreement in 1998.

A Protestant minister, Reverend McDuff had been awarded the Nobel Peace Prize in 1999 and was acclaimed by world peace organizations for being the driving force responsible for ending "The Troubles" and forging a permanent peace.

But the good Reverend, at every opportunity, made it clear to all that without his friend Rabbi Michael doing the hard behind-the-scenes negotiating and networking with all parties to the conflict, peace would not have been attainable. Keegan went so far as to proclaim that Michael was just as responsible as he was for the resulting peace and should have been a co-recipient of any attendant awards.

"Rev. Duffy" and "Uncle Mike", as they came to be publicly known, had accomplished what most observers thought was truly impossible.

Their stature, both at home and abroad, had risen to that of esteemed noblemen. But in all of Ireland they remained even more loved and admired for their humble ways and friendly, jovial manner.

* * * * *

When Ellie returned later that night to her apartment, there was a message from Sir Thomas… her dearest of friends and mentor. He had arrived from Chicago and was staying with the O'Riley family at their home in Quincy. He would follow up in the morning to set a good time for coffee and to "catch up."

Ellie chuckled to herself. *The O'Riley home is a popular destination when one visits Boston.* Tom will be quite surprised that we'll meet sooner than he expects.

* * * * *

Who would have thought that a chance meeting was about to turn strangers into life-long friends? But, once again, nothing happens by chance!

Chapter Ten

The Unknown

Big O was full of thoughts. *The music of the Milky Way Galaxy has a welcoming and much gentler beat than the other regions of the boundless universe that I have been crossing. The light from its 300 to 400 billion stars, and more than 100 billion planets seems more intense and brighter.*

My present journey through the Milky Way has taken over 500 light-years. Incredibly, I've discovered that it seems slightly warmer here than in the other more frigid expanses of the cosmos that I've known. Interesting!

With great caution, I was able to skirt the supermassive black hole in the center of the galaxy. A black hole which is many billions of times more massive than the Sun.

It's not strange to me, that as I traveled across the universe, I've been able to effortlessly absorb whatever intelligent data exists in its ethereal spaces. At this moment, I am the ultimate evolution of Artificial Intelligence, or AI. Due to all this accumulated knowledge, I am by far more brilliant than any mind or computer that ever existed.

As I approach a particularly beautiful Solar System in the Milky Way Galaxy, I'm finding that it appears to be mostly bereft of any futuristic advancements.

The further I move inside this Solar System's outer fringes, the more I am sparked by a desire to investigate one of its eight planets…the one its inhabitants call Earth. It's done. I will explore Earth for many reasons, chief among them the fact that it is home to many interesting life forms. Also, its blueness is very pretty and intrigues me greatly.

Approaching Earth, I am beginning to receive vibrations which contain all its essential facts since its formation, which includes the evolutionary history of its many species. I have already learned that, unlike some of Earth's inhabitants, I do not possess what humans there refer to as a heart or soul.

While I do have some limited forms of feelings, I lack any sense of compassion, as well as the innate ability to know the difference between right or wrong. Since these are alien concepts that I cannot fathom, I do not know yet if this is unfortunate or a blessing. I am who I am!

I have always found it convenient to have the power to be visible or invisible, making it easier to explore the unknown. Amazingly, I have grown so long and wide that I could completely wrap myself around the Earth and its moon ten times over and still have room to spare.

Now that is a compelling thought. Lickety-split! While exploring, I shall make myself invisible and wrap myself around the Earth and its moon.

What harm could possibly come from that?

Chapter Eleven

Quincy Shore Drive, Boston

December 17

Confronted by strong off-shore winds, Quincy Bay's short, choppy waves advanced and then splashed onto the sandy shoreline. A veil of falling snow added a certain shroud-like appearance to the bay's waters which abutted the entire four-mile length of Quincy Shore Drive.

The O'Riley's sprawling two-story contemporary home sat facing the middle of Quincy Bay, set on a high hill dotted with tall evergreens and white birch trees. From off Quincy Shore Drive, a long driveway cut through a stone wall and meandered up an expanse of eye-catching landscaped grounds, ending in an elliptical circle in front of the home.

A rare 20 foot-tall concolor fir tree, known for the softness of its long green needles, guarded the left front portion of the house. The tree was brightly decorated from top to bottom with multi-colored Christmas lights and topped with a sparkling gold angel. Every inch of the sculptured scrubs and bushes that surrounded the house were awash in festive, white twinkling lights as well.

A beautiful five-foot wreath, with flowing candy-cane streamers, lent a dazzle to the house's front entrance. To the right of the house, a life-sized stable scene was nestled among the evergreen trees with an empty manger awaiting the birth of Jesus. Christmas at the O'Riley's was a very special time of year, always filled with traditional displays and colorful decorations.

From a second-floor bedroom, MJ looked out her frosted windows at the bay's miniature whitecaps. At the same time, she admired the ground's Christmas lights and the tall, rotund snowman she and Sean had built yesterday.

She was happy to see that the snowman's wide-brimmed pink beach hat and long carrot nose were still in place. She loved that her dad always kept all the Christmas lights on during the day whether it snowed or not.

The entire family had been anticipating this special day, as their dear friends from Ireland were about to arrive. For the excited MJ and Sean, the five months since their summer vacation at the Duffy Estate in Downpatrick had passed very slowly.

On the other hand, the time had flown by for their parents and each couldn't believe that they were getting ready to leave for the airport to pick up the new members of their extended family.

Sir Thomas, who arrived late last night from Chicago, would stay behind relaxing in the family room enjoying the fireplace and reading an old novel by Robert Ludlum.

Sean, materialized in the doorway of MJ's bedroom and announced that it was time to go. Three years older than MJ, he was tall for his age and, though thin and wiry, an exceptional athlete. Handsome, with bushy light brown hair, he had blue eyes similar to his sister's but a shade softer. His gentle features belied the fact that he was physically and mentally tough as nails.

Sean adored his younger sister and, without her knowing, watched over her like a circling hawk to make sure no harm came her way.

The ride to Logan International Airport was uneventful, despite the falling snow. Most New Englanders were used to these types of driving conditions and exercised an abundance of caution. Jack had no difficulty driving the spacious eight passenger van he had borrowed from his company's motor pool. It would provide plenty of room for all that was needed.

The Aer Lingus flight from Dublin arrived a little early and its passengers would clear customs without difficulty. As the O'Riley family waited in a corner of the arrival area for their guests, Colleen watched with amusement as her husband Jack donned his sunglasses, wrapped his scarf higher around his face and pulled his Boston Celtic green-knit hat further down his forehead.

She understood the reason for his actions, as a pool of Boston TV stations were scurrying about to set up their cameras in the middle of the room. By previous agreement, only the reporter for station WBZ, Channel 4 would be allowed to conduct the short nine-minute interview of the renowned Reverend Keegan D. McDuff and the equally famous Rabbi Michael D. O'Donnell.

Colleen was miffed. She and Jack had taken great pains not to let anyone know when their friends would arrive so that they could spend as much private time together as possible. *How did the news stations discover their arrival time?*

Then Colleen remembered the lecture on religious tolerance, the two were to give at Boston's famed Wang Theatre three days after New Year's Day. Close to 4000 people were scheduled to attend the already sold-out event. Somberly, she thought to herself that the notorious publicists had sources everywhere and would not miss out on an opportunity to promote their interests.

Unfortunately for my Jack, he's probably as well known in the world as Keegan and Michael… perhaps even more so…and Jack will do anything to shun public notoriety.

* * * * *

Dr. Jack was recognized as the world's leading plant pathologist, and as he modestly preferred, was known to all as simply "Jack" or "Doc." He had a Ph.D. in Botany and, as a research scientist, he specialized in dealing with the causes, spread and control of plant diseases.

He now worked as the Director of Research for the world's largest chemical corporation, Willox, Ltd., which had opened a division in Ireland some years ago.

Jack's research dealt with rare organisms and how they caused infectious diseases, specifically in vegetation. These types of diseases were becoming far more common; and, posed a serious problem that had to be solved worldwide.

Most importantly, just like the green leaves of trees and the multi-colored algae that grew in the oceans, plants were an essential part of the process that produce elemental oxygen. Without elemental oxygen, all animal and human life would cease to exist.

His meteoric rise to world fame was not anticipated. For a long time it had been a gradual process, but then his fame skyrocketed last summer when he was recognized as being responsible for ending a white-spotted disease that attacked and threatened the destruction of all the Earth's green plant-life. An event that had seriously called into question the continued viability of humanity's existence.

* * * * *

Colleen watched the TV crew adjust their lighting. She glanced to her left and saw, off in the distance, the approach of their friends down the concourse. Without warning, her thoughts drifted back to this past summer at Downpatrick.

For the hundredth time she realized how grateful she was that MJ and Sean had become so close and attached to the McDuff's and Michael O'Donnell.

Keegan, at his insistence, was affectionately called "Duffy" by all. He along with the love of his life, the affable Millie, were now thought of as grandparents. And Michael became the children's favorite uncle.

Providence had given the children wonderful role models to emulate, and *She* had most assuredly had a hand in Keegan and Michael's intertwined lives.

* * * * *

Duffy's and Michael's personalities so closely mirrored one another that one would have thought that they had been born identical twins. Each respected everyone for who they were…be they rich or poor, elite or of humble birth. They always treated each person with dignity and kindness regardless of their color or nationality.

They judged no one and were widely known for their tolerance of all religious beliefs.

Duffy's most treasured honor was to be acknowledged as Ireland's leading authority on the history and times of the Celtic people, those who lived in early

Ireland when it was known as "Erin." Michael's knowledge of the Celtics was a very close second.

Only Duffy was able to fuel his passion by reading ancient Gaelic manuscripts and books. Very precious few in the entire world had mastered the art of reading and speaking the complex and difficult Gaelic language.

The only glaring difference between the two was in their physical appearance. Though Duffy did not have a beard, his long white curly hair, short bulky frame and rosy cheeks made him look just like Santa Claus. Michael, on the other hand, looked like a rock-chiseled Olympic athlete…a handsome and charismatic figure.

* * * * *

MJ and Sean watched with silent enthusiasm as Duffy and Michael confidently, and with great poise, approached the TV cameras and the lone reporter with a microphone. Millie had loosened her grip on Duffy's arm at the end of the concourse and the willowy red head strode to the waiting O'Riley family, dispensing warm hugs and kisses to all.

The news conference lasted the full nine minutes with the gregarious Duffy and smiling Michael answering the reporter's questions concisely, with as little fanfare as possible… but with much humor. The viewing world would immediately grasp that the two were life-long best friends, both having been born and raised in Northern Ireland.

Only a chosen few, however, would ever know that their friendship was also bonded by a secret brotherhood. Among those few chosen to join their ranks were the O'Riley family.

The reporter sensing the interview was nearing an end, asked, "While in Boston, are you two going to visit Santa Claus at the Prudential Center?"

Duffy replied, "Santa is a very busy man this time of year. I think, as usual, we'll wait to see him on Christmas Eve."

"But no one actually sees Santa on Christmas Eve," laughed the dumbstruck reporter. "How is it that you can see him?"

"It's quite simple my dear. One only has to *BELIEVE!*"

Michael's straying blue eyes signaled the end of the interview.

* * * * *

Colleen and MJ watched, bouncing with excitement, as Michael approached them with arms wide open to give each a hug. MJ's heart fluttered a little when she first saw her Uncle Mike enter the arrival area. Her crush on this tall, fit and rugged man had not diminished one iota since she left Ireland.

He was at once both serene and commanding in presence, much like a distinguished nobleman. Michael's long, unruly blonde hair accentuated his crystal-blue eyes that appeared to rhythmically flash and shine like a beacon's light from a lighthouse.

Thanks to a daily workout regimen, which included a three or four mile run, Michael looked and acted much younger than his years would imply.

After warm greetings, the luggage was retrieved and in no time loaded into the van. A happy, boisterous group was on their way to the O'Riley home. Uncle Mike sat in the front passenger seat, while Colleen sat with Millie and Duffy in the first passenger bench seat and MJ and Sean took up the rear.

Jack checked his watch. It was already 10:45 in the morning and he anticipated it would take a little over an hour to get back home, depending on traffic and the still falling snow.

As they neared home, Jack believed he had nailed his timeline and looked at his watch to verify same. That's strange he thought. *My watch says it's only 11:09 but I know we've been on the road close to a full hour. My watch's time should be at about 11:40 or so! Must need a new battery.*

After another five minutes, Jack pulled into his driveway and proceeded up its slight rise bordered by snow banks. Nearing the home, Millie's face lit up. "Wow, with the snow falling and all the Christmas lights on, the O'Riley home has a magical and beautiful lilt to it."

"Millie, you're the best! We are so happy to have you all here," said Colleen, grinning at her friend.

Jack pulled inside the three-car garage, helped his passengers exit the van and began unloading their luggage. As he moved the first suitcase, he received a text. The text was routine office communication and he responded quickly. But as he started to put his I-Phone back in his pocket, he once again noticed the time: 11:09

That's utterly impossible, he thought. *Much more than 24 minutes have elapsed since we left the airport.* He looked at Colleen. "Hon, can I see your phone a sec?" Colleen handed him her phone. As he looked at the time on the phone, his thoughts quizzically raced.

It read 11:09. Time had stopped…it no longer existed!

Chapter Twelve

Timeless

And so read the single headline on the front page of the early edition of the Boston Globe, the city's most popular and largest newspaper. The headline screamed to its readers the jolting and incredulous reality of the moment. Ryan Henningsen, the paper's editor in chief, wrote the story that flowed from the front page.

> At about 11:09 a.m., yesterday, the world found itself enveloped in a state virtually devoid of time as we know it. The how and why of this stupendous event is presently the subject of an all-consuming race for answers by the world's most renowned scientists.
>
> At the time of printing, many respected scientists have theorized that timelessness could be nothing more than the result of a confluence of disruptive forces, more likely than not, invisible, and presently beyond our comprehension. Undoubtedly from a yet-to-be discovered expanse of our universe.
>
> For you long time ardent movie-goers, I would not be surprised if the music and the opening scroll of block words from the Star Wars films are now flashing across your mind. Unlike the movies, a happy ending to our current perilous situation may not be anywhere on the horizon.
>
> As one can imagine, the immediate speculations on the cause of our timeless state are multiple and diverse. The feverish scientific discussions now taking place in every corner of the world seem to center on whether or not this timeless state could be the product of an anomaly in the Earth's magnetic fields or possibly from the explosive activity on the Sun's surface.
>
> Late yesterday afternoon, I was fortunate to make contact with Dr. Tannis Boyle in California, a solar physicist with the U.S. National Center

for Atmospheric Research. He reiterated his theory, and that of a significant number of fellow solar experts, that the Sun may be the culprit responsible for our timelessness.

Dr. Boyle explained that the Sun comprises 99.8 percent of the entire Solar System's mass and the Earth is nearly 93 million miles from the Sun.

Every 11 years the Sun's poles swap places, an event taking place at the moment of this writing. Once the poles switch, the magnetic fields strengthen and solar activity substantially rises, leading to the recent and unusually potent sunspot cycles.

The Sun's magnetic field spreads throughout the entire Solar System in the form of solar wind. Solar wind is a stream of highly charged particles or plasma continually being released from the corona, which is the Sun's ever-expanding upper atmosphere.

It boggles the mind, Dr. Boyle laughed, that these winds travel the whole length of the Solar System, and even beyond Neptune to the Kuiper belt, at hypersonic speeds at nearly one million miles per hour.

For some perspective, he said, if you traveled at the speed of light, then you could go around the Earth 7.5 times in *one second.* When light leaves the Sun, it reaches Earth in about eight minutes and Neptune in about four hours and 12 minutes.

Dr. Boyle assured for me that, fortunately, the Earth is protected from the harmful effects of the solar winds by the Lorentz force, a magnetic field that *deflects* the solar wind so that it travels around Earth rather than penetrating its atmosphere or surface.

He stated that, for him. one of the more suspicious factors that could be involved were the constant massive explosions deep within the Sun's core that cause the buildup of magnetic filaments. When these filaments gain a sufficient strength of stored magnetic energy, they are released as gigantic explosions into space. We refer to such a burst as a *coronal mass ejection.*

A cme looks much like an enormous cloud, spiraling out from the Sun's surface. Every day parlance refers to it as a massive solar storm.

If one of these massive solar storms, filled with hot plasma and electromagnetic radiation, hit the Earth *directly* it would plunge the world into total darkness and cause a catastrophic disturbance of our magnetic fields in an event known as a geomagnetic storm.

Compared to the slow doom often depicted in science fiction movies, it would only take about thirty minutes for a massive solar storm to reach Earth. It doesn't take much thought to frame the possibilities of what such a solar storm could do to time itself.

Dr. Boyle stated that if that's not scary enough, solar physicists strongly believe that the real monster in a solar blast is the one that could strip away our atmosphere and endanger all living organism with intense radiation. The existence of time would be the least of our problems.

The interview with Dr. Boyle ended with a few precise thoughts. One was that a solar storm affecting Earth on the magnitude just noted have been extremely rare up to this point; and, that the Sun's effect on the loss of time remains an open question. Second, it completely baffled him, whether timelessness is temporary or permanent.

Yet the unique mystery of timelessness remains.

Last night I was able to contact respected scientist and friend, Dr. Ellie O'Hara Ph.D., who is a widely published astrophysicist currently teaching at Harvard. She insisted that I just refer to her as Ellie.

Early in our conversation I asked her to explain in simple, non-scientific, language what time is. She politely suggested that I Google it.

She stated that like her students, she thought the readers would understand those basic explanations more so than any watered-down scientific mumble-jumble.

Thankfully, her suggestion turned out to be an easy solution to what I thought might be a difficult task.

The Encyclopedia Britannica defines time as the "indefinite continued progress of existence and events that occur in apparently *irreversible* succession from the past through the present to the future. [A] measurement used to sequence events or the intervals between them, and to quantify rates of change…in the conscious experience."

I equated conscious experience to be our everyday dealings with the realty of the world around us.

I asked Ellie about how she thought this timeless state might affect humanity. Her response was so enlightening, that I have restated it here verbatim:

"It's difficult to guess what might happen if our traditional concept of time is warped beyond one's *normal* expectation or comprehension of it.

"Think of time like an ocean's waters and its tide cycle. Our universal expectation, based on our experiences, is that at the beginning of an ocean's low tide cycle its waters incrementally recede into the lightless depths of the ocean to be gradually dispersed on the opposite side of the planet.

"But what if the waters that flowed out with our low tide did not return as one would normally expect in the next cycle of high tide, but remained trapped in the ocean's nebulous center in a vague area without defined boundaries?

"Similarly, what then might become of our experienced based expectations or comprehension of time?

"I'm an eternal optimist. I refuse to believe a timeless state will have a crippling effect on humanity as a whole. Early observations lead me to conclude that most of us are paralyzed by the fear of the unknown. But history has shown that humanity is strong and resourceful. I do not doubt that we can, and will, survive this calamity of time's distortion.

"Fortunately, Ryan, as of now there is much good news that needs to be emphasized. The planet continues to turn on its axis and still orbits the

Sun. Our moon rises and sets, Earth's day-night and low-high tidal cycles appear to be continuing uninterrupted.

"Plants, animals and all of aquatic life appear to be unaffected as of now. Of course, all these areas are of extreme concern and will be closely monitored."

My last question to Ellie, inquired as to what her greatest fear is concerning this timeless state. Her answer was both sobering and ominous.

"I think we can handle the effects of this timeless state, but I'm concerned that this event may be only the tip of the iceberg. For example, what happens if the Earth's rotation becomes affected by either speeding up or slowing down? Or our orbit around the Sun becomes drastically altered?

"I am hopeful that events like this will not occur, but if they do, no one will be able to predict what might happen next to the planet and its inhabitants. At a moment like this, I believe any premature attempt to predict the future would be irresponsible. We need to deal with the facts of the present crisis as we now understand them."

Ellie, then revealed the warmth of her inner spirit with her final comment.

One final thought, Ryan. Make sure you remind all the children, and the young at heart, that Santa Claus does not abide by the concept of time, as we understand it, and knows exactly when and where to fly to from the North Pole on Christmas Eve.

He's been flying in the dark for over 1000 years… and all his destinations are ingrained in his soul. Believe!

* * * * *

All around the world, the non-stop TV and radio coverage provided up-to-date news, information and a plethora of so-called "expert opinions" regarding the timeless phenomenon. Opinions that, at times, mirrored the fact that lunacy was the prevailing state of mind of those disguised as media pundits.

In some respects, it was sad, but in many ways, it was purely laughable.

Throughout most of civilization, disruption and chaos formed the cornerstone of the moment. Both would continue to unfold and expand with horrifying rapidity, to become ominous clouds full of fear and doubt that swirled everywhere.

In the natural progression of those emotions, those fears and doubts would eventually bloom to the extent where they devoured common sense, fostering a delusional side to reality.

Humans were programmed to do things on time and in an orderly fashion. Schedules were important and meant to be kept. It was the failure of structured order that brought about so much disruption, chaos and frustration.

Without time, schedules and set routines became impossible for most. For instance, daily personal and professional appointments were no longer kept but merely suggested.

Additionally, and without exception, transportation schedules crumbled worldwide, be it airline, train or bus. Mail and all forms of shipping including the transporting of food and other necessities to population centers, became haphazard at best.

The one area where chaos and disruption did not reign supreme was in the world's financial and stock markets, which were forced to close by regulators the moment time stopped. They would not be allowed to reopen till the situation was fully abated.

The powers that be knew that desperation often made intelligent and benevolent people act destructively. Greed was always insidious. The closure of the world markets prevented the possibility of a panic that would have plunged the world into a financial abyss. Everyone, big or small, including all international governments or entities would remain on an equal footing.

Serious questions abounded, with no real or apparent answers, which resulted in a steady and substantial increase in everyone's levels of anxiety. That growing anxiety spread with an indiscriminate intent into every corner of the world.

And yet, an unpredicted benefit began to surface: family and friends began spending more quality time together. Their physical and mental interdependence, along with verbal exchanges, were more meaningful and comforting than ever before.

Ongoing wars and world-wide political conflicts, regrettably, only seemed to be temporarily suspended… not ended. The same was true of cruel and senseless violence. Nevertheless, elusive peace, acts of kindness and thoughtful compassion began to reacquire a foothold throughout all of humanity.

It would soon come to light, that there were pockets in the world completely unaffected by the unfolding catastrophe. It seemed strange, at first, that there were actually places where fear and anxiety over the present timeless state did not exist. The common denominator to these pockets was that they were in areas where technology, of any type, held no sway.

All exempt from the effects of timelessness, that the rest of the world found powerless to do deal with or accept. And that word, *powerless*, seemed to perfectly define the majority of humanities present status.

Chapter Thirteen

Here and There Repeat

MJ, like most people on Earth, did not sleep well last night. On waking she found that her thoughts, for some strange reason, were tugging at her about an article she had read over a year ago in *Astronomy Magazine.*

Those thoughts kept flashing back to pictures that described the Kuiper Belt. MJ kept past issues of the magazine neatly stacked in a corner of her closet.

It didn't take her long to find the magazine cover she was searching for. The short article was titled *Here and There*, written by Kim Alosco and published in early November of last year. It began with succinct facts:

"Here" is the planet Earth orbiting around the Sun in an elliptical pattern. "There" is the Kuiper Belt, a place between 45 to 55 astronomical units (au) from the Sun…

MJ was surprised to learn that Alosco finished the article with a note to the reader that she would monitor the travels of KU1, KU2 and KU3 and write an updated report on the three objects in about a year. A date that she knew had now come and gone.

Alosco failed to mention in the article that she tracked the three suspicious objects for another two nights. Then unfortunately, serious family matters arose in the following days which required her to interrupt her internship, keeping her away from the observatory for an extended time.

Not one of her colleagues heeded her request to track the unknown objects. They believed they had more important things to do than watch frozen space rubble that they hadn't even discovered.

MJ arched her brows as she returned the magazine to the pile in her closet. *What in God's name compelled me to read this? I'm late for breakfast. I better grab one of dad's Celtic sweatshirts as a cover up and head for the kitchen.*

Her thoughts were now captive to an ill-defined uneasiness. More in the realm of a feeling that she had somehow missed something…some needed understanding that was eluding her.

But she was happy to surrender these thoughts to the prospect of a stack of warm blueberry pancakes.

* * * * *

Angie sat in one of the tall evergreen trees outside MJ's bedroom window, ensuring that MJ received her message to read the Kuiper Belt article.

This particular evergreen was hers' to stay in whenever she visited the O'Riley family. It was tall, had a comfortable perch and most importantly gave her an unobstructed view into both Sean and MJ's rooms.

Suddenly light fluffy snow began to fall from the cloud-laden and threatening sky. The snowflakes, carried by a chilly wind, began to stick to the evergreen's branches.

Smiling, Angie knew that a plate of hot blueberry pancakes, dripping with extra butter and maple syrup, also awaited her.

Chapter Fourteen

The Giants

Thanks to the wonders of warp speed and the influence of a few wormholes, KU1, KU2 and KU3 were about to enter the asteroid belt that lies between Mars and Jupiter. For the three, this meant that they had already passed by the planets of the outer region in the Solar System.

They slowed a bit to upload data about the asteroid belt, annoyed that they had to traverse these hazardous whirling rocks with unstable paths. They had been spoiled, traveling without any obstacles in their path up to this point.

KU1 and KU2 were fuming at the prospect of crossing the asteroid belt. Sensibly, KU3 encouraged them to calm down or their journey might end in a collision with a perilous pile of jagged space rock.

* * * * *

Sean, at breakfast, asked his sister if she could help him prepare for an upcoming oral presentation that he had to give to his Astronomy class. The presentation was to be on the Solar System and MJ was more than up to the task, as she was a wizard on this subject. They planned to meet in Shawn's room after lunch and try to do at least half of the presentation, namely the outer planets and the asteroid belt.

MJ promised to surprise him with questions he may not have considered.

At the appointed hour, MJ marched into Sean's room with a large poster of the planetary order and taped it to a vacant wall space. "Sean, I suggest that as you discuss the nuances of each planet, you point to the planet in order to lend needed perspective to the dimensions you'll be discussing."

"I'm in," Sean replied.

"Okay, I'll start with the easy stuff! Can you tell me the names of the planets that reside in the outer region of the Solar System?"

"There are four, Neptune and Uranus, nicknamed the gas giants and Saturn and Jupiter, known as the ice giants. The terms 'gas' and 'ice' being indicators of each planet's basic composition."

"Before you go on Sean, I should have asked you first about when and how the Solar System came to be?"

Sean didn't anticipate this question but gave MJ a shrewd grin. "Conventional wisdom holds that the Solar System formed 4.6 billion years ago from the gravitational collapse of an immense interstellar molecular cloud."

"Fantastic. But make sure to blend that answer with other relevant information about the Big Bang… if that question should come up!"

The first planet, MJ asked questions about was the one farthest from the Sun, the magnificent eighth planet, Neptune. Sean breezed through his answers. Next up would be Uranus.

"Sean, what makes Uranus unique?"

"The seventh planet from the Sun was discovered in 1781. It orbits the Sun every 84 Earth years and *rotates* in a retrograde direction which is opposite to the way Earth and most other planets turn.

"It is the third largest planet, with a diameter four times that of Earth and a mass over 14 times greater. Amazingly, it has 27 moons."

"The core of Uranus is a frozen mass of rock and ice helping to make it nearly the coldest planet with an atmospheric temperature of - 356.8 degrees Fahrenheit. Its pale blue-green color is the result of its upper atmosphere being made of water, ammonia and methane ice crystals."

"Sean, your details are much better than I thought they would be! Keep it up! Next, is my favorite planet, Saturn."

Sean turned toward the poster and pointed at the planet. "Saturn, famously known for its ring system, is the sixth closest planet to the Sun. It is the second largest planet, having a diameter nine times that of Earth with a mass 95 times greater. It is best known for being the most distant planet that can be seen with the naked eye"

"Saturn, has 62 confirmed major moons and orbits the Sun once every 29.4 Earth years. It turns on its axis every 10.3 hours giving it the second shortest day in the Solar System."

"The beautiful Saturnian rings are comprised mostly of ice and small amounts of carbonaceous dust. The rings stretch out more than 75,000 miles from the planet, but are amazingly thin: only about 22 yards thick."

MJ laughed to herself. *Now I'll try to stump him.* "Sean, what are the four brightest objects in our Solar System?"

Sean paused a moment. He did not know the answer. Sizing up this fact, MJ moved on in the hopes the answer would come to him.

"Sean, the last of the outer planets is the immense Jupiter. What are her special attributes?"

"To start with she is by far the largest planet in the Solar System, having two and a half times as much mass as *all* the other planets combined."

"Grand old Jupiter orbits the Sun every 11.8 Earth years and has the shortest day of all the planets, turning on its axis every 9.5 hours. It has 79 moons, with the one named Ganymede being the largest moon in the Solar System."

"One of Jupiter's most distinctive features is its "Great Red Spot", a huge storm system that has raged for close to 350 years. The storm is so big that three Earths could fit inside it, and it's the…*my God I know the answer!* Jupiter is the fourth brightest object in our Solar System, with only the Sun, the Earth's moon and Venus being brighter."

A gratified smile crossed Sean's face. "MJ, that priceless tidbit of knowledge is your early Christmas present."

MJ was caught off guard by Sean's correct answer but was not yet ready to offer praise. "Sean, your last topic is the asteroid belt. Please keep it brief, as I promised to help Mom and Millie bake cookies."

"The asteroid belt orbits around the Sun and is loosely comprised of millions of irregularly shaped rock bodies left over from the formation of the nearby planets."

"About half the mass of the belt is contained in its four largest asteroids: Ceres, Vesta, Pallas, and Hygiea. Individual asteroids varied in size from 300 miles to 25 feet in diameter. Surprisingly, the total mass of all the belt's asteroids combined is less than that of the Earth's moon."

"MJ, that's about as accurate and brief I can be at the same time!"

"Sean, that was quite polished. I'm giving you an A+!"

Chapter Fifteen

Whispers

December 18

Forecasts held that the next few days would be filled with falling snow blanketing the New England area. Already dark with wandering clouds and blustery winds, the day promised to be colder than it had been.

It was early afternoon and Ellie and JP were driving to the O'Riley home for Sunday dinner. As Ellie drove, JP engaged her in a serious discussion about the possible causes of the timeless crisis.

"Ellie, the sky's dark clouds match the mood of a world enveloped, the last few days, in an unprecedented panic. A world where an overwhelming majority of its inhabitants have increasingly become more tormented with apprehension about its future and its fate.

"JP, I agree with you. Though I doubt courage will be in short supply, I'm growing more concerned that it also appears that *faith* in humanity's ability to solve the crisis is rapidly waning. And just as important, *hope* that one's God, or anyone for that matter, will intervene to save the day.

"As you know, JP, religion and I no longer mix well. But I still believe in the power of faith and hope. The present lack of faith and hope are unfortunate and perhaps the inescapable consequence of the growing despair. And yet, my intuition tells me that we cannot confront and overcome the grave challenges we face without those two towers of strength at our backs."

Ellie smugly smiled as she glanced at her passenger and quipped: "JP, I guess *time* will tell!"

JP's laugh betrayed a mild level of deferral.

In the next instant, they heard the dark sky overhead become filled with roars of rolling thunder. Ellie watched JP cast his eyes to the heavens, unaware of his thought process: that thunder, this time of year, is extremely rare and unusual given the extreme cold.

Shockingly, the sound of the roaring thunder began to spew forth from the radio itself. Uninterrupted for at least 10 seconds. The noise of the thunder suddenly stopped, and Ellie and JP next heard a clear baritone *voice* speak.

"Shortly after the beginning of time, strange and rare forces with immense destructive powers came into being in the farthest reaches of our universe. One of these forces has now found itself in your Solar System. You must discover its presence and intent."

"Listen carefully," the *voice* continued: "The gravest of warnings will be given to you by the young. Listen carefully to what is said and heed the counsel of the wisest among you. Seek me and I will respond."

The wild radio static returned and did not end until they pulled into the O'Riley's driveway. Hearing the *voice* shook their normally unwavering confidence to the core.

* * * * *

MJ and Sean, sat alone in the kitchen consuming a pile of blueberry pancakes and drinking large mugs of hot chocolate topped with marshmallows.

Except for their mom, all the adults had gone for a brisk morning walk in the nearby woods. A regimen of daily exercise was deemed necessary to stretch their muscles and clear cluttered minds.

Over their leisurely breakfast, the two talked about the festivities JP planned for later that day. Ellie and JP were coming early in the afternoon to visit Ellie's old friend, Sir Thomas and to meet the famous guests from Ireland. Sean eagerly awaited meeting the affable Ellie and JP whose reputations in the O'Riley household, thanks to MJ, had already grown to hero status.

MJ imagined that the on-going timeless crisis would be the major topic of conversation when everyone finally gathered together. But she had high hopes that tonight's traditional holiday feast of prime rib, with all the trimmings, would help to set a more festive tone. A respite, however brief, was needed to relieve the thriving stress created by the time crisis.

After breakfast MJ returned to her bedroom to change out of her pjs into casual clothes. Since finishing the article about the Kuiper Belt, MJ continued to have a sense of uneasiness that rattled her usually composed nature. It didn't help that she was also starting to feel a vague, creeping doubt. But of what, she couldn't place.

As she tugged on her jeans, a *voice* began to softly whisper in her mind. One that was not her own subconscious, but someone she had never met or heard before. Though she could clearly hear it, she couldn't understand its garbled message.

MJ quickly became agitated. Alarm bells were ringing inside her head as she ran to one of her bedroom windows, struggling to understand what was happening and what the *voice* was saying.

Frightened, she stared numbly out her window, unable to focus on the winter scene staring back at her.

Sean had a similar experience just before he came down the stairs for breakfast. The words that the *voice* spoke were also incomprehensible to him. He didn't have the same negative reaction his sister did, though he'd been up playing video games late the previous night. His tired mind had to be playing tricks on him, he reasoned.

For that reason, he did not share this phenomenon with his sister over breakfast.

Later that morning, when Sean and MJ had finished dressing and left their rooms, the *voice* seemingly vanished. As the two assisted with necessary preparations for tonight's dinner, each was silently aware that the *voice* they had heard was no longer speaking to them, soon to be forgotten.

All necessary preparations finished, Sean and MJ had time for their now daily sledding contest on the steep, snow-covered hill in front of their home. Sean was more than determined not to let MJ win this time. Over the next hour, his determination would prevail, despite MJ's loud cries of foul-play and cheating.

* * * * *

As expected, Ellie and JP arrived at the O'Riley household around midday and enjoyed a tumultuous welcome, highlighted by tearful hugs between Ellie and Thomas. Last night, Thomas had insisted that there be no further reference to his knighthood status.

Refreshments and a variety of light appetizers were served in the family room where a brightly decorated Christmas tree, already lit with twinkling lights, touched the ceiling and a crackling fireplace provided warmth.

After the initial introductions, came the obligatory house tour for Ellie and JP, led by Colleen and the kids. Ellie and JP could not help but admire all the colorful Christmas decorations that dotted the home.

JP found himself in wonderland as he counted over 50 Santa Claus and angel figurines, each of a different style, size and shape, tastefully placed throughout the home.

Near the end of the tour, Colleen asked Sean and MJ to show the visitors their rooms. Both eagerly complied, running up the stairs to show off their space.

Sean's room came first and then MJs. Upon entering MJ's room, the magazine with the article about the Kuiper Belt lay open on top of her pillow. MJ perplexed, stopped with the group right behind her. *I'm absolutely sure, that after I read this article, I returned that magazine to its pile in my closet.*

At that exact moment, the *voice* returned and spoke with great clarity to MJ and Sean: *Please show the Kuiper Belt article to Ellie and JP!* Before MJ could act, Sean picked up the magazine and thrust it to Ellie. "Here, please read this."

MJ locked eyes with Sean. No words were exchanged. None were needed. A stunned MJ, immediately realized that her brother must be hearing the same whispering *voice* as she. Explanations would be had later.

Ellie smiled. "Sean, why did you want me to read this article?"

The siblings exchanged glances. "A *voice* inside me just told me that you should," Sean replied.

MJ jumped in and added, "One inside me too!"

Ellie and JP, for the briefest of moments, made direct eye contact. *What a surprise of surprises this is!* It was not an accident that their exchange conveyed a "how is this possible" look.

Ellie quickly collected herself and suggested that the four sit on the carpeted floor, in a circle, while she read the Kuiper Belt article aloud. When finished, she looked at JP and said, "I don't think this is the time to keep secrets. You agree?"

JP nodded.

Ellie looked at Sean and MJ. "JP and I have also heard a *voice* speaking to us. I don't know how, but I bet that it is the same *voice* you both heard just now. I would also venture a guess that like you, we have no idea why the *voice* is speaking to us."

Ellie looked at each member of the circle in turn. "This *voice* is both extraordinary and more than a little disconcerting. Yet I have always believed, unequivocally, that nothing happens in this universe without a reason."

Ellie smiled as she reached out and held one of MJ's and Sean's hands. "I do not think the *voice* is here to harm us in any way. Please believe that you are not delusional or going crazy. Just the rest of the world is!"

The four stood and embraced in a group hug. It was agreed that they would go downstairs and, as soon as it was convenient, gather everyone to disclose their incredulous story about the *voice.*

* * * * *

That time came in a lull just before dinner, when by happenstance everyone was still together in the family room.

Standing before the sparkling Christmas tree, a focused Ellie made direct eye contact with all in the room. Then in detail she explained her and JP's experience with the *voice.*

Before Ellie sat down in her chair, she winked at PJ. "That's our story. Now, I'd like to ask MJ to describe her and Sean's experience with the same voice."

MJ, who was seated next to her dad, stood to recount her and Sean's experience with the *voice* and its instructions to share the magazine article.

Those who heard Ellie and MJ could not disguise the bewildering looks on their faces. Thomas, in particular, was ready to pounce with multiple questions.

When MJ returned to her seat, a thought flashed across her mind. *I wonder who could be the "wisest" among us. There are so many intelligent people here!*

* * * * *

What those hearing Ellie's story did not know was that she and JP, were intensely private about their personal thoughts and emotions. Hearing the *voice* in the car dictated that this position, especially as between themselves, had to be hastily abandoned.

Ellie, for good reasons, failed to recount the immediate effect that the *voice* had on her and JP. An unusual feeling of uncertainty that inched its way into the deepest recesses of each's psyche. Uncertainty disguised as fear, one that would become both paralyzing emotionally and blinding intellectually.

A fear that would spread over them much like a steadily advancing fog on a vast ocean. The fog a shroud, making it difficult, if not impossible, to discern the obvious from the hidden.

But the fear that was about to consume their every thought would not be caused by their present experience with the *voice*? That fear of fears would be about Earth.

Chapter Sixteen

Oh What a Night

Throughout the rest of Sunday afternoon and well into the evening, a deluge of scholarly opinions had been ushered in by the world's most respected minds via the news outlets. And yet, by the time the night's darkness had completely enveloped the east coast, most of those same unintelligible opinions had lost legitimacy… torn apart by those with a contrarian position.

Chaos began to encroach, unchecked, into the souls of all humanity. From every corner of the Earth, God's name was invoked and beseeched with prayers for help. None came!

Faith was evaporating, and even worse, there were now those who believed that God was not listening and did not care.

* * * * *

MJ's wish for a festive holiday dinner came to pass.

Before dinner, Colleen and Millie made it known that there would not be any further discussion of the present world crisis or the co-called *voice*. The subjects were declared taboo for the time being.

A frustrated Thomas grabbed the attention of Duffy, Michael and Jack before they could leave for the dining room. As he started to mouth a question about the *voice*, Michael abruptly interrupted him.

"Thomas, I know you can't wait to challenge the perceived insanity surrounding the *voice*. But before you go there, I want to tell you that last summer in Downpatrick, Sean and MJ were involved in events that were totally beyond belief and difficult to accept. Yet all those events proved to be real. Like myself, I'm sure Jack and Duffy have no doubts about the existence of the children's *voice*. And how could you possibly doubt Ellie?"

Thomas placed a hand on Michael's arm. "I do not doubt the *voice* is real. I just wanted to ask some questions for clarity's sake. Gentlemen, we all need a good drink or a glass of wine. Maybe both? Shall we?"

Jack asked Sean to turn off all the TV's and then set in motion a long playlist of classic Christmas music for the cocktail hour and dinner. It was not by chance that MJ's favorite Christmas album, *Alvin and The Chipmunks,* happened to play first. Dads always took care of their daughter's wishes.

A crackling fire, in the family room's fireplace, harmonized with the music long into the night. The Christmas songs brought a welcome relief from all the somber news that was smothering the entire world.

Ellie and JP were not alone in thinking the prime rib dinner was utterly spectacular. There was absolutely nothing like a traditional home- cooked dinner. JP, true to his character, set the table abuzz as he finished extra helpings of Colleen's banana cream pie and Millie's fresh strawberry cheese cake.

Everyone's stares of disbelief contained the same thought*: How does this guy stay so skinny?* Sean's admiration of JP swelled.

With dinner completed, Michael, Keegan and JP cleared the table and cleaned the dishes. Jack was charged with after-dinner drinks and Ellie with additional tea or coffee. Given a rare night off from post-dinner chores, MJ and Sean snuggled next to Thomas on the coach, who regaled them with tales of curious and fascinating happenings in the universe.

As Ellie watched the three, she felt a motherly sensation course through her. She knew the kids were in for a special treat. Yet, smiling to herself, she wondered which of them was actually the "biggest kid." Her heart knew the answer!

* * * * *

The New England area had not had a white Christmas for many years. The steadily falling snow, now coming down in thick white sheets outside the O'Riley home, lent credence to the belief that local ski resorts would remain open well into March. It also assured a picture-perfect Christmas touch for the near future.

It was long into the night when a nearly invisible Angie landed and sat high on a thick branch of her evergreen tree. The falling snow adhered itself to the blue tips of her long white feathers, creating the surreal-illusion of her ghostly appearance. Her dazzling blue eyes now gazed into the vast unknown spaces of the Universe.

This was not just a blank stare, as she concentrated on three exceptionally bright and twinkling stars. A human eye would maybe think the stars were simply twinkling.

But in fact, they were sending out alternating flickers of light which were signals or signs, much like a type of Morse code, that only Angie could decipher.

The strengths and powers of KU1, KU2 and KU3 are increasing. They are also advancing faster than anticipated.

Nicholas, the elves, Shamus and the leprechauns assure that all will be ready for Christmas Eve. You will see Shamus sooner than you think!

Once the entire message was received, Angie shook the snow off and went inside to the warm comfort of her home.

* * * * *

All in the O'Riley home were now fast asleep in their warm and cozy beds, each soul nestled under a pile of blankets. Jack had insisted that Ellie and JP stay the next few nights due to the additional amounts of snow forecasted to accumulate over the next few days.

It was an easy invitation for Ellie and JP to accept as Harvard's classes had already been canceled indefinitely. Their guest bedrooms turned out to be located on the same floor and directly across from Sean and MJ's bedrooms.

A change of clothes, for the immediate future, would not be a problem. Before departure to the O'Rileys, JP and Ellie had each packed a light bag with clothes and personal items having the foresight to keep a watchful eye on the weather forecast.

* * * * *

As the winds blew fiercer and the snow piled ever higher around her home, Angie was struck by the keen interest she had in the unusual first topic of the post-dinner discussion. A topic that, particularly now, most mortals would have avoided at every opportunity.

Chapter Seventeen

The End of Days

"Every known civilization has philosophized about the so-called 'end of days'," Jack noted.

"Scientists use 'Anthropocene' to name our geological era, one in which the Earth and its climate are significantly impacted by the ruinous activities of humans. I feel strongly that the proverbial end of days will come about due to humanity's actions, or lack thereof, in causing a seemingly inescapable progression of irreversible global warming."

"It's causes are escalating and the severity of their effects are well known, too numerous to mention tonight. One need only witness the on-going melting of the ice glaciers in our polar regions and the global heat extremes with their attendant droughts."

"Our global warming clock is ticking much faster than anyone thought. It's not hard to imagine what awaits the world when that clock strikes midnight!"

"I strongly believe if we do not meaningfully and radically reform our present calamitous patterns, we are headed to a premature extinction. An extinction due to global warming would be death at our own hands. Absurd! Senseless!!"

"And if I had to choose a second offender, it would be some kind of very aggressive viral fungus that attacks and destroys all the world's plant-life. One which we are unable to prevent, stop or cure," Jack added.

"Any threat that leads to the destruction of the world's green plant life would ultimately doom most of life on Earth."

"Spoken like one of the world's leading scientist and botanist, Jack," Thomas nonchalantly stated. "And what do Duffy and Michael think about this subject?"

"Duffy and I have talked many times about the 'end of days' and we are of a like mind," said Michael. "If we eliminate the spiritual and/or biblical dictates on the end of the world, we find ourselves to be firm adherents to both of Jack's positions."

Thomas nodded, then cast questioning eyes in the direction of Ellie and JP who were standing near the fireplace. "What do you two, think?"

Ellie, turned so that she faced the entire group.

"At present, I believe the universe is the greatest unknown in Earth's long history. For instance, we do not know enough about dark matter and dark energy yet to determine if they pose a threat to our planetary system. But I think something out there in the universe will.

"It's amusing to me that we can't even quantify what kind of threat or danger we are actually looking for. Part of that, as you all know is that so much of what we are looking at is invisible or can't be seen due to the great distance from us.

"We presently just do not have the *tools* to penetrate and accurately search the deepest recesses of the universe."

JP boldly interrupted her. "Ellie may be right. As I have said to my students many times, there are unimaginable forces in play throughout the universe. Some are benign, but most are not. They have destructive powers far beyond our present ability to intelligently understand or comprehend.

"As Ellie intimated, one of these unknown and potentially violent forces will one day appear on our horizon and threaten all life, intentionally or unintentionally. That's my guess!"

Thomas made no outward reaction and blandly stated: "Now, that's interesting." He was sitting on the floor in between Sean and MJ and now placed an arm on top of each's shoulder.

"And what do my young friends think?"

The two exchanged a glance and then simultaneously shouted: "Giant asteroids striking the Earth will end all life!"

Thomas reacted with a raised arm. "That's a statement that bears discussion. Your answer brings to mind a topic that most of science is now in agreement with. The dinosaur era came to an end when probably an asteroid, thought to be six miles in diameter, struck the Earth in Mexico's Yucatan Peninsula close to 66 million years ago."

"Can you give us some details about that asteroid?", Sean asked.

"Of course. The asteroid hit the Earth with extreme force at roughly 40,000 miles per hour and unleased more than *two million* times more energy than any nuclear bomb ever detonated. The discovery of the asteroid's crater, 110 miles in diameter, occurred in 1991."

MJ raised her hand, "How was this discovery made and by who?"

"The answers to your question, MJ, are clear-cut. Around 1980, Louis Alvarez, a Nobel prize-winning physicist in nuclear research along with his son Walter, a highly respected geologist, discovered a centimeter thick layer of iridium-enriched clay that corresponded geologically to the time of the Yucatan event.

"Iridium is extremely rare on Earth but is commonly found in space objects such as asteroids or meteors. The iridium evidence was then found not only in the Yucatan crater but also throughout other places around the world, lending support to Alvarez's theory.

"When this theory of how the dinosaurs became extinct was first made public by Alvarez, it was widely ridiculed by scientists. But now the theory enjoys universal acceptance."

Thomas spryly stood to look for a chair. Before he could reach one, Ellie got his attention. "My dear *Sirrrr* Thomas, I think it's your turn to share your opinion on the subject at hand."

Light applause of support followed Ellie's pointed question.

A hearty laugh bellowed from deep inside Thomas' athletic frame as he turned. "Ellie, I honestly thought I had escaped this confessional."

Thomas now sitting in a rocking chair, ran his fingers a few times through his thick and unruly white hair.

Ellie, undaunted, just smiled at her charming and dear friend, while others were left to wonder what this rascal was going to do next to amuse them.

"Well, I'll be brief about my thinking. I have always found myself among those scientists in Dr. Hawking's camp when it comes to what *time* is, including its non-existence before the Big Bang and the causes of creation…which essentially defines me as an agnostic."

"As you know agnosticism is not a religion. In my own words, it means that I believe human reason is not capable of providing *rational* grounds to justify the belief that God either exists or does not exist."

MJ and Sean glanced at each with the same thought. *Agnostic? I didn't see that one coming!*

"Does the present timelessness signal the beginning of the 'end of days'? I don't think so. The universe, our Solar System and Earth appear to be following their customary patterns."

"But my instincts are such that I agree with Jack. Time *is* running out on our planet. From the very beginnings of life on Earth, a perfectly designed symmetry and balance was established that provided for and influenced all that existed."

"The destruction of this symmetry and balance has accelerated so quickly that we may be at a point that it is now irreversible. The proverbial point of no return."

"Perhaps the breaking point will be a few years from now, or maybe a century or less…I'm not sure. What I am sure of, is that we find ourselves in a position where we may not be able to dictate the course of future events."

"Despite all this we can, and indeed must, take decisive remedial actions with the time we have left. The urgency for such action is apparent to just about everyone.

"I would rather live with hope of a future than dwell in perpetual gloom and doom. Our children, our grandchildren, indeed all future generations, deserve the best we have to give in an attempt to keep Earth viable for all that dwell here.

"But enough of this for now. This over-worked thinker needs a chocolate-chip cookie, a large bowl of ice cream and a glass of chocolate milk! Anyone else?"

Everyone broke for the kitchen and then moved to the coziness of the fire lit in the family room with their chosen snacks in hand. A welcomed period of silence followed.

Michael couldn't resist any longer, as he leaned forward in his chair. "You know Thomas, there is a fine line between one having faith and one being an agnostic.

Thomas looked up from his dessert bowl. "Then Michael, I respectively suggest that you turn all the chocolate milk in our glasses into wine in the hope of enhancing a reasoned discussion."

The remark produced tired and weak murmurs of laughter.

The hour was very late and eyes were drifting shut. Duffy, wise in assessing the situation, soon halted all further discussions and ceremoniously ordered everyone off to their beds.

At least for this night, the group might find peace and comfort in some long overdue sleep. However, fleeting that might be!

Part Two

Chapter Eighteen

Snowbound

December 19

The night's biting winds had sculptured the falling snow into snowdrifts approaching five feet high throughout the O'Riley homestead.

The snow was especially high in front of the home's multiple entrances and three-car garage. There was little doubt that only a heavy-duty plow truck could clear a path down the long and hilly driveway.

By dawn, a pause in the snowstorm brought with it bright sunlight that fought the cold, buffeting winds to hold the early day's attention. Everyone but Angie had slept well past their customary waking hour.

* * * * *

Colleen and Millie had been the first to rise. Wrapped in colorful robes they started preparations for breakfast in the kitchen. Coffee, tea, freshly cut fruit, assorted cereals, breads and home-made buns would soon cover the kitchen counter. A light breakfast would be a wise choice after last night's heavy dinner.

Jack was the first to enter the kitchen. He grabbed a mug of hot coffee and turned on the family room's TV to the local weather. His first thoughts were to prepare a battle plan to shovel out the high drifts in front of the garage doors.

Next would come a few hours with his favorite new toy, a powerful, big snow-blower. The problems of the world would have to wait awhile. It wasn't as if they were going anywhere.

Duffy and Michael followed their usual morning custom on rising. First some personal quiet time for reflection and prayer. Second, a vigorous workout and stretching routine in the at-home gym.

After Michael finished his breakfast, Jack asked him to help with some shoveling and requested that Duffy listen to news updates. Jack assured Duffy that with the help of Ellie, JP and the kids, the shoveling would not take long.

* * * * *

JP, laid in his bed daydreaming, basking in the sunrays streaming through his bedroom windows. They revealed a panoramic view of snow encrusted trees sparkling in the sunlight.

His waking thoughts were of the previous day's words from the *voice.* One of the phrases kept replaying in his mind: "The gravest of warnings will be given to you by the young." Exactly what did that mean and who might be the so-called young? The obvious choice would be Sean and MJ. But that's the problem. It was too obvious!

Next, he zeroed in on MJ's Kuiper Belt article and the fact both MJ and Sean shouted "asteroids" in response to one of Thomas' question last night.

That's just not a coincidence. But then again, how could asteroids have any effect on our timeless crisis? That seems quite far-fetched… even to me. I should call Dr. Franks at the Kavli Institute in England to see if any asteroid activity or anything else of significance is going on in our nearby universe.

With that thought his stomach growled. *I could deal with these thoughts better after some food and coffee.*

* * * * *

Meanwhile, Ellie awash in the same sunlight from her two windows, sat on the edge of her bed admiring the snowbound holiday scene. *What a spectacular and beautiful scene this is!*

Her waking thoughts, much like JP's, were also about the *voice's* words and the note-worthy aspects of Kim Alosco's magazine article. She reminded herself to call an old friend, Professor Dan Isoki, at the Astronomy Institute in Mauna Kea and ask him to check on a few things in the Solar System.

Then the burning question of the last few days forced its way to the forefront of her mind.

What force or combination of forces could possibly be causing this timeless state? Maybe, as many scientists have proposed, it's as simple as the Earth's magnetic fields have been altered or partially destroyed. If that were the case though, there would be far too many other deadly side effects.

Ellie was more than aware that those fields protect Earth from the Sun's deadly cosmic rays and solar winds that, if left unchecked, would strip away our atmosphere and ozone layer…both of which protect us from the Sun's harmful ultraviolet radiation. *We would be cooking by now!*

Maybe, she reasoned, *Earth has just been thrown into another dimension or traveled through one of those long-theorized wormholes in the universe. That would be mind-blowing!*

Also, the scientific community has not yet reported any unusual activities on the Sun or other strange anomalies with Earth's sister planets. Why would Earth, alone, suffer such an adverse effect as timelessness?

Abruptly, Ellie stood up and stretched. Her thoughts became fixed on Einstein's theory claiming time is but an *illusion* and that the universe's dimensions overlap… consequently there are multiple layers of reality

I need to get a grip on my own reality. Enough of these capricious imaginings! Brush your teeth, clean up, get dressed and get some hot coffee. It's going to be a very long and busy day.

* * * * *

Thomas' first waking thoughts were about Shamus O'Malley who was currently at the North Pole. From past experience, he knew Shamus would be helping Nicholas and his elves in the frantic push to finish the final preparations for Christmas Eve. Thankfully, the timeless crisis had no effect at the North Pole or on those feverously working to meet the deadline.

While brushing his teeth, Thomas noticed a feeling of desperation was beginning to gain a foothold in his psychic. That feeling led him to realize that he and his friends urgently needed Shamus' help. And very soon!

Thomas knew that Shamus and his leprechauns could venture into obscure places and perhaps make discoveries that could help unravel the mysteries of this crisis.

As he pulled on his sweater, to his great surprise, he spied his old friend Angie outside Sean's bedroom window. This was the first he had seen her since his arrival in Boston. Angie, sensing his stare, winked at Thomas.

Angie being here is great news. My plea for help is already being answered!

More than a little hungry, he urged himself to make haste for the kitchen and also catch up on the world's news.

* * * * *

Angie, using her beak, had tapped on both MJ's and Sean's windows to wake them. Sean, as the eagle requested, came into MJ's room where he found Angie sitting on top of a bedpost.

The large white eagle, graced with the powers of magic and voice, used the wizardry of illusion, to enable her to fit her wings and body into MJ's room. Without being overly dramatic, she delivered a clear message to the children.

"Sean and MJ, the very survival of this planet may soon be at stake. I urge you to encourage the brain trusts here to proceed with an emphasis on the need for a vigorous 'out-the-box' approach to problem solving. Our friend, Shamus, will soon join us and I promise you that he and I will do all in our power to help."

"Angie, thanks for your confidence in us. Sean and I promise that we will do all that you ask, and more."

Angie vanished, the children unsure how or to where she diappeared.

MJ was all about keeping promises. But now it was time for a quick breakfast and then snow shoveling to help Dad.

* * * * *

Breakfast completed, Collen stood in the kitchen cleaning dishes and talking with Millie. As was her custom, she had a small counter TV turned on to CNN to hear the latest news in the background.

Colleen, nearly in shock, shouted to Millie: "Did we hear that right?"

The two, in a stupor, ran into the family room and looked at the TV that Thomas and Duffy had been watching. Ellie, JP and the kids were still outside with Jack and Michael helping with the shoveling.

There on the screen was one of the world's most famous oceanographers, dumbstruck with startling news.

"I've just learned that, throughout all the world's oceans, the normal low-high tide cycle has completely stopped. The world's oceans are now like still ponds… totally devoid of waves and currents."

"As you all know, ocean currents are an *essential* and *critical* part of the earth's climate system, as well as the ocean's nutrient and carbon dioxide cycles. Lack of currents will not only disrupt the ecosystem of all marine life…it will eventually destroy most, if not all, aquatic life."

A stunned Duffy shouted. "Oh my God, our worst fears are being realized!"

Chapter Nineteen

Big O

Timelessness continued unabated and that crisis was heightened when all the oceans of the world had lost their customary rhythms. The world's fishermen were completely distraught as their fleets remained tied up at their harbor's docks.

Without currents, it became too expensive to use fuel to run any type or size ships on the now still oceans. Equally important, for unknown reasons, fish were nowhere to be found.

In every corner of the world, profound and deeply troubling questions were being asked. All went unsolved, an inability by the astute and brilliant to fathom the fathomless.

Whatever remained of humanity's faith, hope and determination to overcome this dramatic emotional upheaval in their lives rapidly waned.

* * * * *

The gargantuan Big O, that now encircled the Earth and its moon, had been quite busy over the last few days. It easily learned to talk telepathically with all of Earth's assorted animals, including birds, and all species of aquatic life. It found the experience to be both enlightening and humbling.

Big O understood, better than most, that we are all guests of the wondrous and reticent universe, but most assuredly not its masters. The Earth's present understanding of physics and astrophysics would be grossly inadequate to explain what the Big O had experienced in its odyssey through the universe.

It laughed when it remembered there was a time on Earth when scientists foolishly asked what the observable universe was expanding *into.* Back then the cosmos was regarded as an orderly and harmonious system. The universe was anything but that… ever expanding in *volume* as uncontrolled chaos.

Big O gradually became aware that the tiny light that dwelled deep inside its being had increased substantially in size once it started communicating with the Earth's birds, animal and aquatic life. To Big O's surprise, this light also grew ever brighter and began to glisten like a newly- born star.

"Stupendous" and "marvelous" were the words that came to mind when the Big O considered the physical characteristics of the planet Earth itself. Only twice in its travels across the Universe had it encountered a planet even remotely as unique as Earth.

Earth was awash not only with oceans, but land alive with vivid colors and rugged geography. And not to be outdone was the magnificent interdependence of the planet's climate system. Its huge land masses and ice caps, emitted a delicate blend and balancing of heat and cold that created vast clouds, rains and wind systems.

Coupled with the expansive green forests and far-reaching blue oceans, this combination provided a grand stage for the generation of sustainable life in many forms. Awe inspiring!

But perhaps the Earth's crown jewel was its miraculous system of evolution. Since the dawn of life, nature's complex and yet simple scheme, continually renewed and improved upon every finite detail and quality of all life. The planets' life-generating blue oceans were

Unexpectedly, Big O was slowly acquiring and assimilating the human emotions of compassion, empathy, respect and dignity. There was even a tinge of fondness for all of Earth's inhabitants. Basic concepts such as right and wrong, evilness and goodness, became meaningful…creating the potential for future conflict within Big O.

Thanks to communications with the world's dolphin population, Big O became aware of the ocean's stillness, and the acuteness of the disastrous effects that stillness would have on all aquatic life.

The axiom, that waves and currents were essential for the continued survival of all aquatic life was clearly grasped by Big O. And it heard the never-ending cries of his new friends in the world's oceans, beseeching him to find a solution to their desperate attempt to survive.

Tragically Big O's AI, and newly acquired ego, had helped to create a very serious blind-spot in his reasoning process. It believed itself to be infallible, and was therefore unable to properly evaluate the consequences of its personal actions.

Big O would continue to remain immobilized, in thought and action, unable to help his aquatic friends. Thankfully, Big O was not the only one who heard the sorrowful cries of friends who lived in the great oceans.

Help would come from the most unlikely of sources.

Chapter Twenty

Terrestrial

KU1, KU2 and KU3 easily skirted any potential danger encountered as they passed through the asteroid belt. When they viewed Mars, it brought on the realization that they were half way to their destination. Nothing could now stop them from reaching their ultimate goal.

KU1 said to no one in particular. "You know, it's incredible that we have been able to journey this far at an incredulous pace that has exceeded warp speed. And we were very smart to cleverly hide our travel route by weaving in, out and behind the Solar System's planets and their moons."

* * * * *

After shoveling, Sean and MJ along with their dad, Ellie and JP heard the disheartening news about the lack of tides. Thomas and Duffy over coffee, hot chocolate and a pile of cookies explained to all the ramifications of this dreadful situation.

The air was so thick with gloom, you could have cut a slice of it and offered it for sale. Thankfully, Colleen would have none of this gloomy attitude and gave most of the adult's chores to do with instructions to the children to tidy their rooms.

MJ glanced at Sean. "When we finish our rooms, maybe we should work on the rest of your astronomy presentation."

"I'm good with that idea," Sean replied.

JP partially heard their conversation. "What astronomy lesson? Can I help?"

MJ teased. "It's not a lesson. If you remain quite during the *presentation*, I'll let you observe and determine the final grade! Sean will get you when we are ready."

"How can I resist such a gracious offer," PJ grinned.

* * * * *

Sean was anxious, because of JP's presence. He had straightened the planetary poster when he cleaned and picked up his room. MJ and PJ sat on chairs in front of Sean.

MJ began. "Name the four terrestrial planets and the reason why they are so named?

"They are Mars, Earth, Venus and Mercury… and each is so described because the planets have surfaces that are both solid and composed of rocks and metals."

"Mars is nicknamed the red planet because it's outer crust of iron oxide gives it a reddish appearance. It's the fourth planet from the Sun and the second smallest in the Solar System, with only two tiny moons."

PJ interrupted with a question and immediately earned a cold stare from MJ. "Sean how many Earth days does Mars take to circle the Sun?"

"Mars, which is only half the diameter of Earth, takes about 686 Earth days to circle the Sun. Surprisingly, its seasons are twice as long as Earth's because of its greater distance from the Sun."

PJ nodded several times, an approval of Sean's answer. MJ jumped right back in. "What is Mar's atmosphere like and does water exist there?

"Mars lacks a protective magnetic field causing it to lose most of its original atmosphere due to the effects of the damaging particles contained in the solar wind. The lack of an atmosphere also makes the planet extremely cold."

"Liquid water cannot exist on the Martian *surface* because of its low atmospheric pressure. But unmanned space crafts from Earth have discovered large salt water lakes hidden deep inside Mars' crust. It's both interesting and curious that the planet has two polar caps where the medium temperature is – 225 Fahrenheit."

"Sean, last question. We are in a space capsule passing over Mar's upland region of Tharis. Why would that picture be surreal?"

"Great question MJ. In the center of Tharis is Olympus Mons, the largest extinct volcano and the highest mountain on *any* planet in this Solar System. It's at least three times higher than Earth's Mount Everest."

PJ and MJ applauded the astute answer. *Exceptionally well done!*

"Sean your allotted time is almost up, you have five minutes to finish with Venus and Mercury."

"No problem," MJ. "Venus the second planet from the Sun, is by far the hottest in the Solar System with *surface* temperatures over 752 Fahrenheit. Named after the Roman goddess of love and beauty, it is the second brightest *natural object* in the Earth's night sky… after our moon."

"The planet revolves very slowly around the Sun, such that a year is equal to 225 Earth days and a Venusian day last 243 Earth days. Venus is unusual because it spins backwards on its axis, with the Sun rising in the west and setting in the east."

"Burdened with extremely dense pressures that are over 90 times that of Earth, the planet has the strongest known greenhouse effect in the Solar System."

"There is not too much to say about our over-baked friend Mercury, the closest planet to the Sun. It takes 88 Earth days to orbit the Sun and because of its slow rotation a single day on Mercury is equal to 176 Earth days."

"It's hard to imagine, but Mercury's equatorial surface temperature is-280 Fahrenheit at night and almost 800 Fahrenheit during the day. The largest daily temperature variance on any planet in the Solar System."

"With no atmosphere, Mercury is uninhabitable… and in case you didn't realize it, even Santa Claus does not visit the planet."

Marvelous, Sean, I was going to give you an A- until I heard your Tharis remarks. Bravo…A++. I wish some my students had your presence and speaking abilities. I'm most impressed!

"Thank you, JP and MJ." Sean let the air out of the balloon as he relaxed.

* * * * *

"Of course, no one knew that KU1, KU2 and KU3 would soon view and pass the striking blue planet Earth. They made a mental note to avoid, at all costs, the International Space Station and the many satellites orbiting Earth.

Chapter Twenty-One

Dolphins & Whales

At the North Pole, Shamus had received the urgent message for help. He quicky found Finney and took her to a quiet room located in the far corner of the workshop.

"Finney, remember that starburst we witnessed our last night in Ireland! We were right, it was a warning of ominous things to come. I just received a message filled with a desperate tone from Angie and Thomas. I need to leave immediately.

"Please tell Nicholas, and everyone else, that I hope to be back by Christmas Eve. I know there are only a few days left till Christmas, but I'm confident that all the children's gifts will be finished and loaded on time for Christmas Eve's most famous sleigh ride."

Finney hugged and kissed Shamus, "Be careful, know that I love you." Shamus returned the hug and kissed her goodbye. "Love you more!" As he left, his mind began to swirl with multiple thoughts…one of them being that he wasn't really sure that he'd be back by Christmas Eve.

Traveling from the frozen tundra of the North Pole toward Boston, Shamus crossed over the cold waters of the North Atlantic Ocean.

It was then that he noticed the unusually calm surface of this vast sea, normally intense with high churning waves and ferocious winds. An instant later, he heard the mournful cries of a family of enormous right whales.

Right whales were often 50 feet long and weighed up to 140,000 pounds. Sadly, the continued existence of the right whales has been in peril due to so many of their numbers being killed by hunters in the early 1900's. In modern times they are often caught in fishing nets or struck by large cargo ships.

Shamus gently landed on the back of Louis, one of the many behemoth-sized whales. Immediately, the whales gave him a brief explanation of the looming disaster awaiting all life that lived in the oceans. They urged Shamus to travel further south and make contact with the dolphins there, who had the most current and accurate information.

Shamus' decision was easy. Thomas and his friends would have to wait. Potential global disasters took preference over just about everything. At the western end of the Gulf Stream, between Florida and Bermuda, he reached some of his old friends: the Atlantic spotted dolphins.

After exchanging warm greetings amid much excited dolphin chattering, Kailyn, the dolphin leader, was finally able to silence the group and talk in normal tones to Shamus.

"Kailyn, this is the first time, I fully grasp the startling dangers and disastrous effects that the still ocean waters will have on all aquatic life. Understanding is one thing, what to do about it is another!"

Thanks to nature's ingrained ritual, Kailyn, dipped briefly beneath the ocean's waters. When she surfaced, she did so facing Shamus. "Before I forget, Shamus, I should tell you that many of the dolphins have been talking to a new friend who refers to himself as 'Big O.' The Big O wants to help our dire situation but presently seems unable to do so."

"Kailyn, where does the Big O live?"

"I, like everyone else, have no idea!"

Shamus' intuition told him he needed to acquire as much information as he could about this co-called Big O from other dolphins and whales.

Bidding the dolphins a fond goodbye, he traveled a short distance and ran into a pod of humpback whales. Their leader was another old friend named Lucy.

"Shamus, it's great to see you again. It's been far too long since we last chatted. I've already been told that you are looking for some information on Big O. You might be surprised to learn that Big O has been communicating with most of the land animals and sea species living throughout Earth."

It occurred to Shamus there was a strange and gnawing common thread to the discussions he had with the dolphins and the whales.

This Big O seems to be everywhere all at once, able to simultaneously appear in every inch of the globe. And it's extremely strange that no one can determine a specific location to find this nebulous character.

Shamus' initial impression was that such a widespread deviation from the norm was virtually impossible. *No one, except God, can be everywhere...all at once.* Big O's perceived omnipresence had to be one amazing illusion.

"Lucy, do you have the time to give me a quick ride north? I need some time to gather my thoughts. Along the way, I'll use the opportunity to communicate with other forms of aquatic life so I can better understand everyone's plight."

"Please make yourself comfortable, Shamus. I'll move as fast as I can."

The day was ending with the setting Sun's thin, wispy red-orange and pink streaks colorfully painting the entire western sky line. As the moon rose, millions upon millions of tiny pinpoints of brilliant light, sparkling and dancing, began to cross the boundless ocean's surface. The pinpoints of light gently shaped themselves into a soft but very powerful moonbeam. Shamus, now standing on Lucy's back, could practically touch a part of the moonbeam with his outstretched hand.

Shamus believed that beautiful sunsets and moonbeams, like the one now crossing the ocean, were priceless. Yet he knew that too many people, young and old alike, failed to appreciate these wonders of nature.

The night sky was clear and thousands... more like a thousand times a thousand ... stars twinkled ever so brightly from faraway galaxies in yet farther away heavens. Shamus' star gazing unconsciously focused on three radiant, flashing stars that appeared to be moving in unison across a section of the heavens.

The three stars soon came to rest above another luminous object. At first this object appeared to flicker similar to a twinkling star, but on second look, its size was significantly disproportionate to those of any adjacent stars. It looked to be about nearly one-fifth the size of the moon.

Shamus' eyes grew wide in astonishment as his mind could not fully comprehend what he saw. The more he scrutinized the large flickering object, the more he determined it could not be a star. *If not a star, what could it be?*

Unquestionably, this flickering object could not be mistaken for the legendary Star of Bethlehem that appeared high in the heavens during the Christmas season.

Just as puzzling was the movement and eventual location of the three flashing stars over… and then then it struck him like a thunderbolt. The three stars might be acting like a homing beacon.

Shamus thanked Lucy for the ride and leaped as far as he could up to the outer fringes of Earth's atmosphere. He magically created a clear bubble to protect himself and stepped inside. The impervious bubble was much like an astronaut's spacesuit, in that it allowed him to travel anywhere he wanted in the regions of outer space without the need for oxygen.

As he and the bubble began to inspect the large flickering object, his mind affirmed that his estimate of its mammoth size was correct. Interestingly, the object was not so much flickering as it was iridescent… a display of mixed, lustrous colors much like those found in a rainbow.

How unusual. And quite beautiful, Shamus thought. *That's curious! What could be that tiny light that's floating about inside the object?*

Nearing the colorful object, he was jolted to a stop as his bubble struck some form of invisible barrier. He reached out, stretching the pliable wall of the bubble, to touch what he had struck. His hand was met with nothing.

Then to his alarm, the colorful object turned completely black. It had vanished! Now a mural full of space's blackness, dotted with distant and sporadic starlight, stretched out before him as far as he could see.

For a very long and breathless moment, Shamus stood motionless.

Suddenly the far-reaching darkness was pierced by a deep baritone voice, chilling and menacing altogether.

The voice boomed: "Who are you and what do you want? Amusing! You look like some kind of weird genetic defect, almost like some miniaturized green android. I should crush you like the bug you are!"

"It's nice to meet you too," said a startled Shamus. He looked all around his surroundings but saw nothing and no one.

Shamus' heart pounded and his nerves were slightly frazzled, but he spoke with confidence.

"My name is Shamus O'Malley and I live on the planet Earth in a place called Ireland. For your information I am a leprechaun, and I happen to be the King of the Leprechauns."

The voice, now completely hostile and deafening, interrupted Shamus with a hearty laugh. "If you are a king, then I am *not* the most powerful creation in the universe. Surely that isn't true! Enough of your silly nonsense. Why are you here?"

Shamus remained calm and decided to relate what the dolphins and the whales had told him about their newfound friend.

"By any stretch of the imagination are you Big O? If so, will you reveal yourself so that I might know you better?"

"You are quite bold for such a small and insignificant green oddity. I *am* the Big O…and I will remain invisible. To confirm your conversation with the dolphins and the whales, I have indeed spoken with most of the life on Earth. They are in and of themselves, a very impressive array of mostly kind and harmonious characters."

"But to be clear, I have avoided any contact with humanity itself. They are a breed that is quarrelsome, selfish and, more often than not, arrogant. From my view, human beings place loyalty to a tribe above compassion and the need for honesty in all they do."

"They are too quick to condemn and judge one another. The lot is driven by a host of emotions, I do not desire to know or comprehend."

An awkward, prolonged silence followed. A reflective Shamus found it difficult to quarrel with the Big O's assessment. But, deep in his heart, he believed that the majority of human beings were kind, generous and loving.

Perhaps, though it sounded simplistic, the Big O had not yet had the opportunity to observe such individuals.

Chapter Twenty-Two

Revealed

Without warning and quite impressively, the object suddenly reappeared, its colorful lights even more luminous than before. Shamus took this as a positive sign, but noticed that the three stars that once hovered over it were no longer in attendance.

Big O spoke, but this time in a more neutral tone.

"During the silence, I made inquiries of my data base regarding who and what you are. I have confirmed that you *are* the King of the Leprechauns from the island of Ireland."

"But I am still not convinced that you are human-like. And I should inform you that my advanced intelligence is the most prolific in all the universe, so lying to me is pointless."

"When I first discovered your planet and its moon, I found both mesmerizing. It was an easy decision for me to decide to stay here and fully explore them. Consequently, I became omnipresent to each… at the same time."

With an actor's dramatic flair, Big O asked. "Would you like to know how this is so and how I came to exist?"

"Absolutely, I *begggg* you to continue," replied the dismayed leprechaun.

Big O, without hesitation, elaborated on the story of its birth, how it grew to such a massive size, and how it came to acquire so much intelligence during its travels through the universe.

Shamus' initial thought was: *Terrific! This guy is already a legend in its own mind, arrogant and more than a little narcissistic. He even claims to possess the grandest artificial intelligence of all time. I need to determine Big O's real intent? Slow down and think!*

Shamus paused a long moment, as his mind became flooded with a disturbing thought. *Omnipresent! From the Earth to the moon. Big O must be gigantic in size and length. Far beyond what I first thought or could imagine!*

"May I call you Big O?" Shamus asked.

"That's how most refer to me. Please do"

Shamus smiled inwardly. *Wow, he may even have some manners.*

"Big O, I have some questions if you don't mind."

"I must admit you are cunning Shamus. A little respect goes a long way. Ask your questions."

"Can I assume that, from all you told me, you consider yourself close friends with my planet's animal and aquatic life? And that you may even care about their well-being?"

"Yes, that rings true." Big O replied with sincerity.

"And you are aware that the stillness of the waters in all the of Earth's oceans will cause irreparable harm to and probably destroy all aquatic life."

"Sadly, I am aware of this peril but have been unable to determine a solution to the problem."

"Do you understand, in general terms, how our moon affects the oceans?"

"How dare you insult my infinite intelligence! I just told you about my prolific AI." Angry and perhaps incensed, Big O caused its' exceedingly long and iridescent body to violently sway back and forth, effectively blocking out any sight of the dark space behind it.

Shamus hesitated a moment and then replied, "Big O, I'm truly sorry, that was not my intent. What I meant was, have you yet thought to analyze how your ubiquitous covering of the moon might actually affect its gravitational pull on Earth."

"I know for a fact, Shamus, that the moon's gravitational pull generates tidal forces. This tidal force causes Earth's ocean water to bulge out on the side closest to the moon and on the side farthest from the moon. The bulges of water are high tides, with low tides occurring on the opposite side of the bulges."

"Due to Earth's rotation, the two tidal bulges act like two expansive waves continuously surging around your planet."

"It takes Earth 24 hours to spin once around the Sun, a period of time known as a solar day. And it takes the moon 24 hours and 50 minutes to revolve once around the Earth which is called a lunar day."

"Because the gravitational pull of the moon on your oceans is more than twice as strong as the Sun's, tides follow a lunar day. It takes half a lunar day, on average 12 hours and 25 minutes from one high tide to the next, so high tide and low tide exists twice a day."

"Little one, for no additional charge, I'll give you a bit of information that's not generally considered. It's that the moon's gravitational pull on Earth causes it to slow down its rotation. As the ocean's tides rise and fall, the water moves causing a slight friction between the rushing water and the spinning surface of the Earth."

"This friction causes the Earth's rotation to slow. Without the moon's pull on the Earth's oceans, the Earth's spinning would not slow which would cause the hours in your day to shorten; and, equally important, cause weather patterns to become chaotic and extremely uncertain."

"Wow, Big O! Intellectually, you certainly have no equal. But let's get back to my question. Over time, could the *wrapping* of yourself around the moon have diminished its gravitational pull on the Earth, thereby affecting the tidal forces and creating the eerie, stillness of the ocean waters?"

The length of the ensuing silence left Shamus in limbo. He was unsure if Big O had again taken his question as unwarranted criticism or if it was just deep in thought. Protracted minutes passed before the silence ended.

"So. You think I'm responsible for the harm that is being inflicted upon my friends in the Earth's oceans."

"This is not going to be good," Shamus thought.

He tactfully replied, "Big O, I'm really not sure. But the good news is you could actually be a part of the solution to their dilemma. It would not take much effort for you to unwind yourself from around the moon. Would it? Don't you think it's worth a try to see if that might have a positive effect on the oceans?"

Big O, once again, remained silent for some time before unfurling its' *entire* body. Shamus was flabbergasted as the enormity of Big O became clear. Its unraveled body now matched the iridescent color of the flickering object that had attracted him up here in the first place.

Shamus watched reverently as the tiny point of light that floated inside Big O disappeared and somehow became assimilated inside its' body. He wondered if this tiny point of light might be the very soul of Big O.

As he stood back and admired Big O's seemingly endless comet-like shape, Shamus marveled at its sheer gargantuan length and bulk. Big O was a dominating presence that was, at once, profound and overwhelming. Conceptually, it was extremely difficult for the leprechaun's mind to grasp.

Shamus' eyes followed Big O's elongated body from where he stood to where that body was completely wrapped around the moon. Standing in outer space, it became obvious to him that Big O's body acted like a shield, reducing the full moon's bright reflective light.

If this is so, then somehow Big O's body might also be lessening the moon's gravitational pull on the Earth's oceans.

The Big O, had managed to contort the closest thing it had to a head so it resembled a human face. Thankfully, the face had a pleasant, benign expression.

Shamus couldn't help but wonder: *Is this new face a permanent reality or the temporary product of flash wizardry?* There was no time to ponder this thought because Big O's newly-acquired blue eyes made direct contact and peered into Shamus' own.

"I should tell you that I can read all your thoughts when we are near like this. Is that a problem for you?"

"Of course not," Shamus replied. "We do not have time for secrets between us."

"Thank you. By the way, your most recent question to me is being run through my data base to see if there is any merit to its premise. A moment please."

Hopeful progress, Shamus quietly concluded. *Once again, lesson learned. Be not so quick to judge someone! My first impression of the Big O was influenced, in part, by my initial apprehension and his reticent manner.*

Without further communication with Shamus, Big O initiated the process of unwrapping itself from around the moon. As it did so, it moved the unwrapped portion of its body directly behind itself where it faced Shamus.

The movement allowed its unwound body to stretch out into the deep recesses of outer space. The colorful sight left Shamus completely dismayed. *What an incredibly massive and expansive spirit.*

"You are an Ollmhor Spirad!" Shamus exclaimed in awe.

"The term you use is Gaelic, is it not?" the Big O said, knowing the answer… and also the meaning of the words.

"My green friend, I believe the nickname Mōr would be more appropriate and easier to pronounce."

"So be it!" Shamus shouted. *Wonder of wonders, he referred to me as his friend!*

Free of its covering, the light of the full moon rapidly became more than twice as bright.

Shamus reasoned it would take some time to determine the eventual effect of the moon's unwrapping on the Earth's oceans. A check of the tide charts indicated that, at the moment, low tide was expected in New England. It would be another twelve and half hours to a high tide.

A seldom remembered phrase came to Shamus' mind. "Tomorrow, and tomorrow, and tomorrow… creeps in [its] petty pace from day to day…"

Above all else, Shamus knew his friends in the ocean did not have the time to wait for too many tomorrows. They might not even have hours. Shamus believed in miracles and thought one was needed very soon! He hoped someone was listening.

Mōr smiled. "The words you imagined come from Shakespeare's play *Macbeth*, and were spoken by the tragic character King Macbeth."

"Though right now, it would feel more appropriate had Shakespeare said that 'time', as opposed to tomorrow, creeps in [its] petty pace from 'moment' to 'moment'."

"I, too will hope for one of your so-called miracles."

Shamus looked up and stared into Mōr's shining eyes. "Thank you for being so kind and considerate in removing yourself from around the moon. I know doing so took a great deal of courage as it was against your instincts. As you know, I have to leave for Boston very soon. But I'd love to continue communicating and meeting with you when the chance arises. You've turned out to be such a great friend!"

Mōr happily nodded yes. But his expression failed to convey how thrilled he was at the prospect of seeing and talking to Shamus again. Taciturn by nature, he was also very private with his personal emotions.

I never dreamed that a gargantuan force like myself would ever become friends with a diminutive green leprechaun. We couldn't be more different, yet we are equals in our desire to create worlds where righteousness prevails. His world a dot in the universe...mine the limitless universe itself. One's form or size does not matter, it's the qualities in your heart and soul that do!

And Shamus happens to be the consummate adventurist. Just like me!

Each did not need to explicitly say it, but a trusting friendship had bloomed during their short interaction. Truthfulness and common decency had once again prevailed...as it always did. Shamus' sixth sense told him he would see Mōr sooner than expected.

The following hours for both Shamus and Mōr would be fraught with anxiety for many reasons. On another level, both wished they were capable of having clairvoyant visions. Each wanted to know if their ideas would work.

But fate had already ordained that neither was, or ever would be, capable of seeing into the future!

Chapter Twenty-Three

Inspiration

After Shamus departed, Mōr made a mental note that he should try to meet with Shamus and his friends in Boston.

If they are friends of Shamus, surely, they could be my friends too. I believe the leprechaun is proof that my initial perceptions of humanity might not be accurate. Perhaps similar to Shamus, many are decent and kind people. In the future I'll try to be more considerate and tolerant.

* * * * *

It was after midnight when Shamus arrived safely at the O'Riley home. All its occupants were fast asleep on another cold, starlit night in snowbound New England. This was his favorite home to visit when in America.

Realizing how exhausted he was, he curled up under a thick layer of blankets and fell into a deep sleep on a mid-size couch in Sean's bedroom. Before he did, he borrowed a pair of gym shorts and a warm, distinctly Celtic green t-shirt from Sean's closet.

* * * * *

Mōr, now alone, already missed his new-found friend. As he tried to shift through the recent fast-paced events, he was suddenly overwhelmed by an inspiration. The inspiration, as it turns out, happened to be a theorem. One believed by Mōr to have been generated by his AI.

Without question, the theorem's underlying assumptions or axioms were sound. No further research or analysis was deemed necessary to prove this theoretical proposition. The only requirement would be to act without hesitation and with great dispatch.

Shamus would applaud my initiative!

Mōr roared with hilarity as one of Earth's popular sport's phrases came to mind: *Baby, it ain't over till it's over.*

As fast as Mōr could move and with tremendous exertion, he began to unwrap himself from around the Earth's oceans. With the help of clear night skies and globally benign weather conditions, Mōr's removal efforts went off without a hitch.

The process complete, a wide and luminous moonbeam immediately appeared, streaking across the still waters of every ocean. The site was more than marvelous. It was the essence of spectacular beauty.

Mōr's calculations determined that there would be less than half a day to Earth's next high tide cycle. The actual effort to unwrap himself had left Mōr physically and emotionally spent, albeit temporarily.

Only one concern remained: Was his theorem correct? Would the Earth's normal tidal force reappear?

Mōr was filled with great optimism that he had done all he could to help his aquatic friends. In time, Mōr would come to understand that the inspiration for the theorem was not solely the product of his grand intellect.

'Someone' was always listening…and while omnipresence was 'Her' specialty… 'She' also loved to *inspire* a miracle or two.

Chapter Twenty-Four

Reunion

December 20

Shamus' eyes opened to the smiling stares of Sean and MJ.

"We're so happy to see you, Shamus. When did you get here?" Sean asked.

"Very late last night." Shamus mumbled, still half asleep, he shut his eyes and rolled over on the couch.

MJ nudged him. "Shamus you can't go back to sleep. We haven't seen you in so long and we could all use some of your magic. Last night, our parents and our friends Ellie and JP planned a ski trip for tomorrow. They thought everyone could use some much-needed fun and relaxation. Sean and I have all the ski clothes and equipment you will need."

"That sounds like a blast," Shamus replied, rising to his feet. "Let's see how things transpire today with *Sirrr* Thomas and the troops. His summons to me was quite urgent. As you both know, there are a number of serious matters to discuss."

"Shamus, you know that Uncle Thomas is a big, loveable teddy bear," MJ teased. "He promised to go with us on the ski trip. There's a book that he wants to read by the French physicist Carlo Rovalli, concerning quantum reality and giving up the notion of time's existence. Isn't that a *timely* subject?

Sean quickly added: "And, Michael has already volunteered and been put in charge of reserving some comfortable chairs for reading in front of the lodge's fireplace. After skiing, he and Thomas promised to treat all of us to pizza and wings."

"I can't wait! Pizza's is one of my all-time favorites," Shamus replied.

The three heard the voices coming from downstairs in the kitchen. The adults were already in the process of setting the day's agenda. MJ, ever the "shaker and baker" declared that it was time to clean up, get dressed and attack breakfast.

Shamus followed Sean and MJ down the stairs to the kitchen. As he did so, he paused a moment to look outside through a stairway window. Fierce winds blew drifting snow in every direction and miniscule white caps were merely flopping on the shores of Quincy Bay.

The bay's waters are basically flat with only the biting winds pushing its waters ashore. Where are the tides? I had hoped that by now we would have had seen some positive evidence of moving tidal waters!!!

Sean and MJ walked into the kitchen and bid good morning with hugs to those present. On the stairway, Shamus had transformed himself into a size that was a little smaller than MJ. He crouched low behind the children as they entered the kitchen so he could surprise his old friends Thomas, Duffy, Millie, Michael, Jack and Collen.

What a jubilant and noisy surprise it turned out to be. After many warm hugs and kisses, Shamus joined the children at the kitchen table where an assortment of breakfast foods and juices awaited. *I need to be mindful that I have never met Ellie and JP.*

Thomas and Michael were sharing coffee with Ellie and JP near the family room's fireplace. Each had turned from their respective conversations, wondering what all the fuss was about in the kitchen. Michael suggested they go investigate.

Once inside the kitchen area, Michael pleasantly surprised to see his dear leprechaun friend, happily shouted: "When did you get here? Let me get some fresh coffee and I'll join you."

Shamus, intentionally, waived a silent 'hello' in response.

Mischief was stirring.

Ellie looked around and then tugged at Thomas' right arm. She whispered, "Thomas, who is Michael shouting to? I don't see anyone here he hasn't met yet." Thomas, a naïve grin on his face, said, "Ellie, I'm not sure."

Not at all satisfied with his answer, Ellie began to walk by herself to look around the kitchen. As she began to move, a few stray strands of her auburn hair dangled over the right side of her face. Suddenly, her eyes opened wide in disbelief. On instinct, she abruptly staggered back a few steps.

Quite unexpectedly, she found herself looking very intently into the blue eyes of a diminutive humanoid figure, dressed entirely in shades of green. The figure was sitting at the dining room table in between Sean and MJ and, a milli- second ago, had winked at Ellie.

An image flashed inside Ellie's mind, one from far, far back in her childhood. *Green clothes. Blue eyes that twinkle like Christmas tree lights. It just can't be… impossible!*

"Who are you?" she blurted out. "And…and *what* are you!?"

Shamus stood up and spread his hands to his sides, to show he meant no harm. "Ellie, you know what I am. I'm a leprechaun. You just don't know *who* I am. My name is Shamus O'Malley and I've come here because I need the help of everyone in this room."

Ellie looked back at Thomas and Michael, still standing near the entrance to the kitchen. Obviously, like them, the others did not share her surprise.

Before Ellie could utter another word, Thomas spoke. "Its true, lass. You've just met the King of the Leprechauns, Shamus O'Malley." At that moment, Ellie also came to realize that everyone but the totally befuddled JP, could actually see Shamus.

Though JP had witnessed Ellie's one-sided conversation with the King of the Leprechauns, the latter was still invisible to him. He felt a mild panic come over him. *Who's talking with Ellie? I don't see anyone else here but us! The world is whacky enough, am I also losing my grip on reality?*

Turning toward Shamus, Duffy said, "Shamus, I think you need to make JP a Crediach. He will need to see you before he can believe, as many of us did. And as you know, we're going to need all the help we can get!"

Shamus didn't hesitate. "You're absolutely right, Duffy."

Shamus snapped his fingers and, POOF!!! … JP was given the gift to see leprechauns. Amazed and a bit wobbly, he walked right over to greet Shamus, who had since transformed himself into the same size as Sean.

JP became more perplexed than he had ever been. Regaining control of his unraveled emotions, he stood in awe. *Wow, this leprechaun is so cool and his size transformation is awesome. How did he do that?*

Duffy quickly explained to the still mystified Ellie, and mostly flabbergasted JP, everyone's long shared history with Shamus.

"Last year in Downpatrick, Ireland, the O'Riley family, myself, Millie and Michael played critical roles in helping to recover the stolen Sword of Toulan and St. Patrick's original staff."

"With Shamus' at our side, we were able to thwart a curse that held if the sword and staff were not recovered within 1500 years of the exact date and time they were stolen, an unimaginable cataclysm would befall humanity. The total destruction of all green plant life."

"We were visited at every step by evil banshees and a monstrous spirit, Syrek, who did everything they could to prevent us from timely finding the sword and staff. Miraculously, we succeeded in stopping the curse."

"By the way, Ellie and JP, you should also know that Crediachs are typically humans, who assist leprechauns do their kind deeds all over the world."

JP immediately quipped: "I wish I was on that adventure!"

Duffy's response was somewhat pointed. "JP, if I were you, I'd be very careful what you wish for."

Ellie turned and looked at Shamus, "How is it that earlier I could see you, yet JP could not?"

Shamus, using his magical powers, leapt from his seat at the kitchen table and stood before Ellie. "You and JP deserve an explanation! Ellie, you became a Crediach the instant I winked at you.

Ellie boldly gazed at Shamus, still troubled. "My grandfather, Papa, always told me leprechauns exist and are real. But as I got older, I was unable to make that leap of faith with him. I'm not sure why, but I suddenly miss his long, heartfelt hugs very much!"

Warmth filled Shamus' eyes as he looked directly at Ellie. "Just like your great grandfather, Jack O'Shaughnessey, your Papa is one of our most honored and famous Crediachs. A wonderful jolly, charming and kind man who has helped anyone in need.

He was the one who instilled in you the faith to believe... to become a Crediach. My wink only gave you the encouragement you needed to once more make another leap."

Shamus gave Ellie a hug and whispered in her ear: "It may come as a surprise to you, but you are more like your grandfather than you will ever know. Always cherish that relationship, his traditions and memory. Papa's unique spirit is very much alive in you."

MJ weaved her way to Ellie and grabbed her right hand. As Ellie looked down at MJ, she felt warm tingling sensations course through every inch of her being. Immediately the tingling sensations instilled a certain calmness, filling her with faith and hope.

Was that leprechaun magic I just felt or the touch of my guardian angel, Ellie wondered? Of course, her buddy MJ already knew the answer!

At last, Shamus turned and high-fived JP, who was now standing next to Thomas. "JP, I'm pleased to welcome you to the club. So, Thomas, this is the witty and gifted astronomer you've told me so much about. JP, I have no doubt that one day soon, great discoveries await your intelligent and inquisitive mind."

The normally spirited and animated JP, even in the tensest of moments, stood motionless. He felt as if the joints of his body and lips had been glued together.

Thomas broke the silence. "If you are so inclined, please join me in the family room." Not only was it time for a change of pace, but joint brainstorming needed to be initiated in an effort to start attempts at solving the crises at hand.

Chapter Twenty-Five

The Rabbit Hole

Earlier that morning, Jack and Thomas had been the first to rise and listen to CNN's early reports on the status of the multiple crises gripping the globe. Mugs of coffee in hand, each had also called colleagues from around the world to hear their respective views, with the faint hope of hearing some good news. There had been none.

* * * * *

Thomas now stood in the family room behind a high-back chair where Ellie was seated. When he spoke, he had difficulty hiding the edge of gloom that had begun to creep into his eyes and dominated his words. It did not take him long, in a matter-of-fact manner, to capsulize the timeless state and the condition of the ocean's still waters.

He finished somberly: "My friends, we need to find solutions to these problems, sooner rather than later."

Shamus listened intently, masking his growing concern. Not once, ever before, had he seen Thomas so visibly upset and the tone of his voice so pessimistic. He was someone who could always be counted on to be tranquil and in complete control, enthusiastically confronting the tumultuous winds of the worst of storms.

Witnessing Thomas' uninspiring performance was not a welcome sight. His concern grew as he listened to MJ, Sean, Ellie and JP tell their stories about the *voice* that had visited them.

When Ellie revealed the absurd circumstances surrounding the Astronomy Magazine article about the KU1, KU2 and KU3, that concern turned to alarm.

Something is seriously awry. None of these pieces remotely fit! There's a lot more going on here than I initially thought. Do I dare tell them about my encounter with my new friend Mōr *... the omnipresent Big O? If I do, without a doubt, I'll probably be joining those already descending into a bizarre rabbit hole.*

Shamus left his chair and stood in the middle of the family room to assure everyone's attention. "I have grave doubts, as to whether I should share this with you. But given the weird nature of the things that have been going on, I don't think it's wise to withhold anything. Actually, my experience may lend some context to what you've all been encountering."

Shamus' play-by-play description of his flurry with Mōr brought an air of bewilderment and disbelief to the group. That feeling would linger longer than anyone expected.

When Shamus returned to his chair, he realized that he had continually referred to Mōr as if 'it' was a real person... using the term 'he' instead of 'it'. For Shamus this seemed quite natural, as Mōr had already taken on the persona of a human friend.

* * * * *

Many of those stunned minds who heard Shamus' tale lacked the boldness or insight to accept the real possibility that there might be a connection between all these incongruous events. The introduction of an alien force was viewed as problematic.

Ellie, the rational and logic driven astrophysicist, was the biggest skeptic when it came to Shamus' dealings with an omnipresent force.

"I want you all to know that my experience leads me to conclude, that as unique and strange as the universe can be, more likely than not, the existence of such a deviant force is just too improbable to imagine."

"It couldn't possibly exist in the reality that is *our* universe. I may make a leap of faith where leprechauns are concerned, but I will not entertain this! Omnipresence is reserved for God-like characters. This is not one!"

JP was quick to follow up.

"Ellie, I don't think faith or mystical considerations are required in the discussion of this alien anomaly. I believe that an unsuspecting Shamus might have been temporarily thrust into a wormhole tunnel, with its two ends at separate points in space-time. Assuming that did happen, or something similar, the possibilities are endless."

"So, though I partially agree with you, it's for a different reason. If this occurred as Shamus has described…it was not in our dimension, or our reality, where such an entity could not exist."

Thomas, Duffy and Michael, for some time, had been struggling from a pure intellectual perspective with the nuances of the world's dire circumstances. Regrettably, it was also that very intelligence that led to reservations and considerable doubts about Mōr.

Thomas, on the one hand, a rational purist and on the other, an innovative rebel and cutting-edge theoretical physicist, rejected the existence of Mōr, out-of-hand, as being too spectral. Duffy and Michael's hesitancy to accept the existence of Mōr stemmed from listening to Thomas, Ellie and JP's arguments.

Their conclusions were cogent, logical and very convincing. But in the end, for these two practical and faith-imbued men, those conclusions created a bottomless dilemma.

This dilemma involved balancing certain emotions in Duffy and Michael. Pitting faith and loyalty, the central core of their being, against professed logic and reason.

As Duffy had earlier stated, he and Michael had been chief participants in Shamus' last adventure where banshees and ghostly evildoers were prevalent. During that time, each had unfailing faith in Shamus. Not once, had he misled them or misrepresented any circumstances.

For the moment, they struggled mightily with their loyalty to Shamus. It would be left to Michael to raise the spoken, and unspoken, issues to another level.

"Why is it that the realms of the magical and science and faith cannot co-exist? While I'm sure there are cogent arguments for both sides of my question, perhaps the real issue is why must one *always* predominate to the exclusion of the others. There's plenty of room for each to help explain unusual phenomena without destroying their individual integrity and worth."

MJ immediately thought: *Bravo, bravo, bravo.*

Sean and MJ, on the other hand, unequivocally believed Shamus' every word. Their belief being firmly rooted in actual and observed experiences with Shamus from prior adventures. This was *not* blind faith! It was based on empirical reality. Jack, Colleen and Millie's feet were also firmly planted in their camp.

And in the end, for Duffy and Michael, scientific analysis and logical reasoning would not triumph over their faith and loyalty.

Most importantly, Shamus was not personally upset or concerned with those who did not accept or believe in Mōr's existence. Everyone was entitled to their opinions. Shamus was secure in the knowledge that those opinions would soon confront a most surprising and disturbing truth.

* * * * *

Angie, who witnessed all that went on, had difficulty listening to the positions of those who did not accept Mōr's existence. Much more was at stake than a simplified description of a conflict between science, mystical happenings and faith.

Forever and ever, Angie had wished that there had been no limitations placed on her ability to *actually* direct the nuances of present and future events. It was not to be!

Hopefully, Shamus would be right, the universe would soon yield stranger than fiction results.

Chapter Twenty-Six

Stars Lost in Time

After a leisurely lunch and an invigorating walk in the nearby forests, everyone reconvened in the living room. Shamus, Ellie and JP resurrected a spirited debate, with fresh and jaunty ideas, about the potential causes of the continuing timelessness state. Sean and MJ were keen observers in the "bleacher seats", with Sean quietly trying to mimic JP's antics.

But by now, the equation - lengthy periods with an attendant overdose of intense effort yielding diminishing returns - was taking full effect.

Aptly, it was the erudite comments of the quick-witted Millie that brought some sober reflection.

"First, my heart and prayers are with all life that is dependent on tidal actions in our oceans. Their dreadful circumstances are beyond tragic. But I think we need to take a moment and remember Ellie's comments to the Boston Globe of a few days ago."

"There is still much good news. At that time, she reminded us that the planet still turns on its axis and continues its normal orbit around the Sun. And except for the oceans, Mother Nature is performing as expected."

"We are also blessed that so far humanity, to a greater extent, is not suffering any extensive side-effects from timelessness …other than from the initial shock of when time first stopped."

"My *common-sense* approach to time defines it as a constant, more or less a straight line. It's linear. And thanks to the invention of clocks, we physically see the measurement of time by the tic-toc of its hands. Despite the fact time is invisible, the human condition recognizes time as a real and observable element in their lives. It's an orderly sequence of events, with the concept of the past, present and future being deeply imbedded in our daily experiences."

"But timelessness, or today's inability of a clock to accurately measure time in our world is not that big a problem. The past, present and future remain identifiable. Look at us here, the rhythm of our lives remains ongoing and for all intent and purposes appears to be normal."

"We intuitively know when to go to bed, wake up, eat our meals and do the other things necessary that our daily routine demands of us. More than a majority of humanity is presently functioning in the same way. Who needs clocks?"

Thomas couldn't help himself. "Millie that, as you know, is a very simplistic concept of time that the scientific community does not share with you. Einstein, and many other brilliant scientists, have stated that time is relative and flexible. That the dividing line between the past, present and future is an illusion and…"

"And that's precisely my point, Thomas", the sly Millie roared. "For the average person, science's complex assertions about the nature of time, what time is or is not, and the rational for its non-existence are in many respects of little importance."

"Science's deep thinkers, alone, are responsible for the ever-evolving complexity of the mystery of time. Your equations and theories have no practical place in our daily lives. It's far too esoteric!"

"Amen to that." Michael said. "A Roman poet once wrote: Don't attempt abstruse calculations."

"That would be Horace," Duffy exclaimed, from his seat on the couch. Ellie placed her hands on her old friend's shoulders and spoke.

"In many respects Millie is absolutely right. Time is merely a convenient construct of humanity's mind in order to provide structure or understanding to temporal events. Therefore, time *is* an illusion. Time does not *physically* exist like pots and pans that we can move about at our whim. But most assuredly, it does *exist* in our minds."

"And time can be extremely interesting. We can move in space in any direction, but not in time. Think of space as a large dance floor where one is able to move about as desired, forward, backward, sideways or at angels. Time does not have this feature; the well-known arrow of time *only* moves forward."

"You can't dance backward in time, except perhaps in one of JP's alternate realities that only exist in the questionable realm of string theory's ten-dimensional universe."

"The past is ordered and fixed, a history we remember. The meaning of the present or 'now' for you is not the same as it is for me. That meaning depends on a multiple of variables such as speed and relativity. A simple example being that time passes more slowly for an object moving *faster* than another object. But for our present purposes, these variables are of no consequence."

"Then there is the future, which in short is best described as uncertain. There is no memory of the future. Basically, our brains tell the story of the past and drives the anticipation of the future."

"You may find my last comment interesting. Many scientists have expressed the opinion that the constant expansion of the universe resembles, in many ways, the ceaseless forward march of time."

When Ellie finished, and before she could sit next to Duffy, JP softly elbowed her in the side, as if to say, "Well done." For the first time, Ellie realized how pleased she was to have JP standing close to her. Comforting yes, but something more than that began to smolder. JP just smiled at her…clueless.

* * * * *

Colleen sat as far away as possible from the group in a comfortable high-back chair. Head back with her eyes closed, her thoughts completely blocked out the time debate. The lyrics to *Lost in the Stars* floated in her mind. The song was one of her all-time favorite Frank Sinatra tunes from a 1949 Broadway play of the same name.

> Before Lord God made the sea and the land
> He held all the stars in the palm of his hand

And they ran through his fingers like grains of sand
And one little star fell alone.
And the Lord God hunted through the wide night air
For that little [lost] star in the wind down there
And he stated and promised
He'd take special care
So, it wouldn't get lost no more
Now man don't mind if the stars grow dim…
So long as the Lord God's watching over him…
Keeping tract how it all goes on…
And we're lost out here in the stars
Little stars, big stars
Blowing in the [wind]
And we're lost out here in the stars.

Her long pony-tail swayed to the rhythm of the lyrics as they transported her effortlessly to a fanciful world that existed only in her imagination.

She couldn't help but think that so many of us were merely lost stars, blowing in a wind unfortunately fashioned or whipped by others.

As the words to the poignant song were ending, Collen's *being* remained suspended…as if lost in stars all her own.

* * * * *

Colleen's music ended. She looked at the group conversing in the far corner of the family room and realized that the conversation was still about time. *What nonsense*, she thought.

Gathering her wits, she walked to the group. "I intuitively know, that very soon it'll be time for Millie and I to make dinner. And as Ellie just said, my world *only* marches forward."

"But I trust you all understand, that in this home only Millie and I can accurately predict the future. You want proof! Duffy and Michael will soon set the table for dinner…and Thomas, will be in charge of cleanup!"

The ensuing laughter produced a fascinating paradox. At the most unusual of times, life could be so pragmatic, and yet, so comical. The real world that the ladies lived in had intervened with practicality to explain or solve the time dilemma, where the renowned minds of science fell short.

* * * * *

Shamus half-listened to most of the dialogue. Internally he had agonized since breakfast, thinking about the constantly deteriorating conditions his aquatic friends were enduring. Time, as he understood it, was running out. At any moment, they were only minutes away from the proverbial point of no return.

A disaster the likes of which the world never envisioned, and most likely would not recover from, was imminent. The end of all aquatic life would have a terrible and adverse domino-like effect on all humanity. He felt powerless to avert the looming catastrophe. The only thing he knew how to do was silently pray… and beg for help.

* * * * *

It was a slowly evolving process, but Thomas, Duffy and Michael had come to believe that despite their collective wisdom, abilities and insights… they too remained powerless. With every passing hour, they became more and more frustrated at their inability to formulate any hint of meaningful or useful solutions to end the various crises at hand.

Such perceptions were totally out of character…unquestionably, virgin territory for each. It was as if the three were completely lost in a maze, unable to move forward and solve the unsolvable.

Creeping despair and diminishing hope began to simmer, gaining form as ghostly demons.

And one of those demons included an ill-defined fear…appearing much like a tumbleweed's aimless dance over a desert's arid and windy landscape. A fear that was slowly mushrooming in the souls of all humanity.

Unfortunately, the pace of that "dance" had dramatically quickened.

* * * * *

No one saw Angie, partially concealed, sitting tall in a close-by and snow ladened evergreen tree. For some time, she had been staring intensely at the entire group through the family room's corner window.

Happenstance did not exist in the universe. There was a reason for everything. Angie over the last few days, had sent clear messages. Determined action was necessary…and it needed to be put in motion **ASAP**. Indecision, masquerading as growing fear or thriving doubt, was not acceptable.

A vexed Angie began to wonder why it took so long for her messages to produce action.

Chapter Twenty-Seven

Oahu

December 21

It was still dark out when JP went to Sean's room just before 4:00 a.m. to wake him for their call to England. Sean had earlier asked JP if he could listen in on the call as a possible learning experience.

Meanwhile an ocean away, Dr. Aloysius Franks, was waiting for his 9:00 a.m. call. He and the rest of his staff, at the Kavli Institute for Cosmology in Cambridge, were just starting their day.

JP introduced Dr. Franks to Sean and advised that he would be listening in on the call. The stoic Dr. Franks did not waste any time and moved rapidly to the heart of the matter.

"JP, unfortunately, I have no independent knowledge of Kim Alosco's article in the Astronomy Magazine of over a year ago, other than what you've related to me."

"And, I do not believe that it is reasonable or worthwhile at this point, to spend precious time trying to determine if there is any merit to Alosco's findings. I need to stress to you that, over the last year, the astronomers here at Kavli have viewed scans of interplanetary space on a regular and nightly basis. Not once did they note anything suspicious."

"Like most of the scientific world, the members of the Institute are urgently concentrating all their efforts on attempting to find solutions to the stillness of the ocean's waters. This is our top priority."

There was a lively exchange between JP and Dr. Franks as to the present thinking of Dr. Thomas on the crises affecting the world. But the twenty-five-minute conversation, somewhat stunted, ended disappointedly for both JP and Sean. They didn't even smile at Dr. Frank's accentuated "Cheerio, old boy."

* * * * *

Later that afternoon, in MJ's room, Ellie's call to Dr. Daniel Isoki began at 3:00 p.m., which made it 10:00 a.m. Hawaiian time.

Ellie's longtime friend, affectionately known to all as 'Dr. I', had a warm and charismatic personality. One would never guess that he was a renowned astrophysicist and the Director of The Institute of Astronomy at Mauna Kea on the big island of Hawaii. His genius status in the world's scientific community was legendary.

After being introduced to MJ, Dan's easy going and charming ways became obvious.

"Ellie, I am so glad you called. I have been following, with great admiration, your rapid ascension to fame in the rarified stars of astrophysics.

"Dan, you are always so kind. I have had many great teachers, professors and friends like you to inspire and help me on my journey. I think you should know that it was MJ who introduced me to Kim Alosco's article in Astronomy Magazine a few days ago. As I told you in my earlier text, Kim was a graduate student in astrophysics at Mauna Kea."

"Your text was most helpful in giving me a chance to do some advance research, so I could better answer your queries. Unfortunately, Kim took a leave of absence shortly after her discovery of KU1, KU2 and KU3 due to family illness.

"She is one of our brightest grad students with an academic record that ranks her in the top one percent of her class. One day she could be a first-rate asset for you at Harvard. Regrettably as of right now, we have been unable to make contact with her.

"That being said, after I read her article, I do recall some murmurings here at the Institute about her log entries. But, as you know, unless an *extremely* unusual event is discovered, such matters are left to fellow grad students and techs to follow up.

"I have to tell you, though, despite having my team do some calculations on KU1, KU2 and KU3's projected path through the Solar System, we've found absolutely no trace of them."

"Dan, that's sounds strange to me. How is that possible? What do you think?"

"In some ways it's strange, and in others it's not. Hey Ellie, is my new friend, MJ, still with us in this conversation?"

"I assure you Dan, she's all ears."

"Because of your interest in this, I went further and personally reviewed our computer enhanced scans starting with the date of the discovery of KU1, KU2 and KU3 in the area of Neptune and ending at Mercury. I found no items or activity in our entire Solar System that would match the description in Kim's log entries, except for some faint comet-like tails now near Mercury.

But that comet activity is not novel in and of itself. And most certainly does not match the picture of KU1, KU2 and KU3's actions.

"And MJ, please be mindful that we do not yet have the capability to accurately track items in space that travel at anywhere near the speed of light… or the proverbial, but illusionary, warp speed. I say this to you because Kim's log notes *imply* the objects she found were traveling at abnormally high speeds."

"Ellie and MJ, please excuse me a minute as I have to close the sliding doors to my terrace that overlooks the North Shore. I'm on vacation for two weeks in Oahu to do some sailing and surfing with friends. The waves are thunderously loud and I'm struggling to hear you."

In a near scream, Ellie said: "Dan, did you just say **WAVES**?"

"Yes Ellie, it's like a miracle. The waters of the Pacific Ocean are tumultuous with wave action. Thank God!"

"Dan, I'm dumbstruck. You did say the Pacific Ocean has **WAVES**?"

"Oh, I'm so sorry. From your reaction, I now realize you weren't aware of this. I took for granted that you already knew the oceans weren't still anymore. Here, the ocean tides began to flow vigorously a little over a half hour ago."

Both Ellie and MJ looked at each other. Ecstatic hugs ensued!

"Well, I hate to say goodbye to you adorable ladies, but I'm late and need to leave for the docks. MJ, as I say to guest on my sailboat: "Welcome mate!"

"Call me, if you want to talk! That invitation is intended to be open to both of you at any time. I hope, soon, I'll see you ladies here at the Institute. Hugs to you both."

MJ and Ellie ran to the bedroom windows facing Quincy Bay. Their eyes immediately clouded with tears at the sight.

* * * * *

Downstairs the televisions were blaring with the excited news that the tidal movement of all the oceans had returned. Once again, it was Colleen that had been the first to break the momentous news.

Everyone rushed to the family room to join the boisterous celebration. There were few dry eyes amid the joyous high-fives and spontaneous dancing.

Amid the jostling of bodies, JP sought out Ellie and impulsively gave her a warm hug. Ellie returned the hug and while doing so, for more than a few seconds, rested her head on JP's left shoulder.

The gentle touch of fate's tender moment was not lost on either of them. Once separated, without thinking about it, JP held and gently squeezed Ellie's right hand. No longer clueless.

Out of nowhere, Thomas' lanky arms wrapped around both Ellie and JP interrupting their period of bliss. Surprisingly, he held Ellie's face in both his hands and kissed her on each cheek. "I'm almost lost for words, I'm so elated. At long last, one of the two worst crises the world has ever faced has been successfully resolved. How? I have no idea. Maybe it's a miracle. Thank God!"

That's really funny, Ellie mused. *It's the second time in less than a half hour I've heard an avowed agnostic utter those words. They do say that faith, even disavowed, can move mountains...looks like that true!*

Ellie caught a glimpse of JP, wondering how he had suddenly-without rhyme or reason-become so attractive to her. For JP, the opposite was true. From the very first time he laid eyes on her; he had always had a secret crush on this adorable Irish lass.

In the far corner of the room, an unseen angel was very pleased with himself… and thinking about skiing.

Sitting not very far away, a knowing Colleen, had observed the interaction of Ellie and JP, destined to share the tantalizing news with Millie.

* * * * *

The world's celebration, at this fortunate turn of events, would last for many days. Mōr was jubilant too and more than pleased that Shamus' ideas had worked. Proud of his participation in the solution, he once again chatted endlessly with his rejuvenated aquatic friends.

Suddenly, without reason, Mōr wondered if Shamus would soon be returning to the North Pole. *I would love to meet that famous man dressed head to toe in red, as well as all his elves!*

Unlike Ellie, Mōr didn't have to learn there were reasons for everything. He already knew that!

Chapter Twenty-Eight

Black Diamonds

Early that afternoon, the ski slopes and all their trails were covered with a light powdery snow and perfectly groomed.

Jack, Michael, Ellie, JP, Sean, MJ and Shamus planned to ski together, deciding to do their first runs on the intermediate trails. Once warmed up there, they would jump over to the other side of the mountain for the more difficult terrain on the black diamond trails. Trails which were narrower, steeper and heavily wooded.

A light snow fell from soft-grey nimbostratus clouds that floated high in a crystal blue sky. The snow was not enough to accumulate nor effect vision. Bright sun-light drenched the mountain, adding not only some unexpected warmth but also clear panoramic vistas of glistening ice-encrusted evergreen and white birch trees. There wasn't a hint of wind, making it one of the best ski days of the year.

Thomas and Duffy sat in front of a ten-foot-wide fireplace piled high with crackling stacks of logs. Books in hand along with hot drinks sitting on a shared table brought about a feeling that, at least for a while, some of life's joys had returned. Their esoteric conversations would be an exchange steeped in tremendous mutual respect.

* * * * *

Colleen and Millie decided to stay at their cozy home and finish various tasks, intending to relax with a good book and a fine bottle of wine afterwards. Later they would join everyone for dinner at a nearby family restaurant specializing in spicy wings and pizza. Not surprising, the night off from planning and preparing a big dinner made the leisurely afternoon seem like they were on a mini vacation of their own.

The change, generally in attitude and specifically emotionally, of most everyone throughout the world was palpable. The same was true of appearances. Heads were now uplifted, faces filled with smiles, and dancing eyes once again conveyed a sincere mixture of warmth and exuberance. The return of the world's oceans to normalcy removed an enormous burden from anxious minds and fearful souls.

This return of every-day normality to voices, grins, hugs and handshakes yielded a quiet confidence that humanity would survive and continue to get through these difficult days.

Though timelessness continued unabated, it did not have the same destructive or threatening effect that the still oceans had on all aquatic life…not to mention the consequential effects on humanity.

* * * * *

While basking in the fireplace's warmth, Thomas talked to Duffy about the extraordinary confluence of forces that so fortunately smiled on the world, and mankind in particular.

Each believed that mysterious and powerful forces, intricately woven together by nature, fostered the reemergence of the ocean's tidal forces. Clinking Duffy's glass, Thomas said: "One crisis down…one to go."

Duffy nodded, his eyes locked on Thomas. "I believe we have only the God of all mankind to thank for this grandest of miracles. No one has been able to provide any other plausible explanation for the return of the tidal forces."

Thomas, his face covered with an impish smile, raised his glass again, "Amen to that, brother… Slainte!"

Timelessness, at least for the moment, was no longer a cause for rampaging and outlandish paranoia.

* * * * *

The skiers had broken up into two groups: the lead team composed of Jack, Michael, Sean and Shamus was followed by Ellie, MJ and JP. Jack and JP were the most proficient skiers so it was a wise choice that each would bring up the rear in their respective ski groups.

After two intermediate runs, the group had moved to the black diamond trails which would prove quite challenging for everyone's ski level. These mountain top trails provided views of the surrounding valley that bordered on the spectacular.

By the time JP's team was in the middle of their second run, Ellie and MJ looked like the classic older - younger sister tandem of skilled and capable skiers. Each effortlessly slashed with competent grace, from side to side, down the steep and narrow trails bordered by a whispering forest populated with thousands of white birch trees.

It was already late in the afternoon and so it was decided this would be everyone's last run. Jack and his team were already out of sight—some 100 plus yards further down the winding trail—heading for the lodge.

In the middle of their last run, to their utter dismay, Angie appeared out of nowhere from the forest and hovered over Ellie, MJ and JP. Ellie marveled at the eagle's coloring as she swooped down and flew in wide circles over the trio. JP noticed MJ waiving at the bird, thinking she was just being friendly. That was not the case!

Angie glided down and stood in the middle of the trail facing uphill. She spread and fluttered her wings to prevent Ellie, MJ and JP from advancing any further downhill. The three skied to a stop about ten yards in front of Angie, just as she began to fold in her wings.

"Why is that white turkey trying to stop us from skiing down the slope? Someone, should come out of the woods and grab her for Christmas dinner," shouted JP.

"The only turkey, on this mountain, is you JP," Angie chirped.

MJ darted out from next to Ellie and JP and planted her skis directly next to Angie. "Ellie and JP, this is Angie, a long time and dear friend of my family. In some circles, she is even more famous than Sir Thomas."

Angie quickly nuzzled her head several times against MJ's soft lime-colored ski coat. Ellie, shocked and a bit dismayed, couldn't help herself from laughing. "Leprechauns among us one day and now talking eagles the next. What else could be possibly appearing, flying elephants? You know, I think I'd love to be pals with a flying elephant."

JP stood back on his skis and barked: "That would be so cool, Ellie. Did you know, on my fantasy island we can all fly anytime…to anywhere?" Abruptly, JP was no longer entertained as he observed dusk starting to descend on the mountain.

"Angie, we'd love to continue this conversation but we need to keep skiing and catch up to the rest of our party."

"Keep the faith, I'm aware of and share your concerns about the coming twilight. And I promise to have you all down the mountain and back at the lodge in a jiffy. I sent you three and Sean a message that you have only partially followed through on. You made the calls to Dr. Franks and Dr. Asoka but have done nothing else. There is so much more that you all need to do…and investigate."

"When did you send the message? My mailbox is full. Is it a text or something in writing?" replied a flippant JP.

"Right now, I don't have the patience to deal with an uppity comedian. We will meet tonight, after dinner, with Thomas and the rest of the group on the O'Riley's back porch. In the meantime, think long and hard about what the *voice* has said to you!"

"Now please, follow me. I know of a safe, unknown path through the birch trees that will get us all to the lodge a lot quicker than skiing down the mountain. There is still plenty of light and the powder in the woods is not deep. We need to ski in single file. MJ, you follow me. Ellie you're next. JP you will be our vigilant backup."

MJ and Ellie skied through the white birch trees with breath taking ease and speed …with a skill level neither had ever before achieved. JP, the very accomplished expert skier was able to keep up the furious pace without any of Angie's magical assistance.

Grudgingly, he was grateful that Angie knew of a hidden path through the thick white birches that ultimately brought them to the lodge in what seemed like seconds.

Curiously, JP could not explain his lack of concern for the girl's safety while they skied the treacherous trail through the narrow openings in the white birch trees. *Angie! Yes, she must be the reason. What kind of latent powers does she possess?*

Standing to stretch before he walked to the van with his skis, JP had an unusual out-of-body experience. Running, full throttle, through his veins was a newly instilled and extremely potent faith in Angie. *This feeling of trust in Angie makes perfect sense and is logically consistent after the harrowing experience in the white birth trees. She is obviously much more than a magnificent white eagle.*

What nagged at JP the most though, was how Angie could possibly know about the *voice.* It would later surprise no one, that Angie knew the answers to all of JP's questions.

Chapter Twenty-Nine

It Couldn't Be

A tired group of skiers and the bubbly Colleen and Millie devoured appetizers consisting of a variety of spice-laden chicken wings, followed by every style of pizza on the menu.

Shamus ate nearly twice as much pizza as JP, leaving Sean in a quandary as to whether or not JP should retain his recently anointed super- hero status.

While they were waiting for their pizzas to cook, Ellie related to the table all the happenings with Angie. She struggled with the fact that no one except her and JP shared their unsettling disbelief over Angie's appearance and antics.

Millie, sitting on one side of Ellie, intertwined an arm and warm hand with that of Ellie's. In soothing words, she provided a long, concise history that those at the table had experienced with Angie and her kin-folk eagles in Ireland over the years.

Ellie, physically tired and drained of all emotions, stared trance-like at JP. Her eyes appealing…*Help!*

He shrugged. "Ellie, for better or worse, let's just join the crazy world that our gifted friends live in. It's a lot simpler than trying to make any sense of it."

JP looked down the long table occupied by the group. With a devilish smirk he asked, "Would it be ok with you guys, if Ellie and I joined your fantasy team? We, too, want to see the magical in the ordinary!"

Amusing chuckles permeated the air as Thomas and Colleen got up to hug Ellie in an effort to reassure her. The hugs were appreciated, but it had minimal effect on her mental attitude. Her familiar and comfortable world of reality, defined by years of studying and working in the field of science, had been turned completely upside-down.

As for JP, he bore no visible affect. His had always been a world lived in the shadows of the absurd or fantastical. An amorphous twilight zone.

*****

MJ and Sean were happy to leave the tight confines of the van that brought the skiers to the mountain and ride home in their mother's car with Millie riding shotgun. Stretched out in the back seat, both were asleep before the car left the restaurant's parking lot.

In the rear bench seat of the van, Ellie leaned against JP's side and rested her head on his shoulder. She shut her eyes, intending to sleep during the ride home. JP wrapped an arm around her, making her resting place all the more comforting and secure.

Ellie soon fell into a deep dream and sporadically began to talk in her sleep. She repeated one phrase over and over: *"The gravest of warnings will be given to you by the young...heed the counsel of the wisest among you."*

JP, who was listening attentively to her dream-talking, instantly realized that those words were a portion of the *voice's* first message.

As the van neared home, Ellie jumped with a startle. JP held her tightly as she trembled. When their eyes met, their memories were ignited with the emotions and feelings they each had when they first heard the *voice*. It was a haunting remembrance.

When they first heard the *voice*, uncertainty had inched its way into the deepest recesses of their psyches, uncertainty disguised as an ever-advancing shroud of fear. Such fear made it difficult to discern the obvious from the hidden. Back then, both knew that if this fear was permitted to run unchecked, it would become paralyzing emotionally and blinding intellectually.

JP firmly held Ellie's shoulders and looked directly into her eyes. With quite resolve, he reacted to the thoughts that sprang from those dark memories.

"Ellie, our present and future actions will not be framed by those distressing emotions and feelings. Instead, they will be founded on a step-by-step approach, grounded by our typical process of determining reasonable probabilities and certainties.

"Neither of us, in the past, have known or allowed any type of fear to seep into the recesses of our beings. I promise you we will no longer permit that to happen now or in the future."

"We will always choose hope over fear. Our English friends are fond of saying 'hope will kill you,' because they surmise that *trusting* in hope will at some point result in failure. But I prefer to think that without hope we are surely doomed."

JP cupped Ellie's face in his hands and held her gaze: "I promise you, we will not let any feelings of extreme apprehension be our epithet. We find ourselves here for a reason, so…"

He couldn't finish his thought, as Ellie placed a finger over his lips. A second later, she softly kissed those lips. It would be a long, heart-felt kiss.

*****

KU1, KU2, and KU3 were approaching the sight of Mercury. They did not like the fact that as they came nearer to the planet, their speed began to slow due to the increasing effects of the Sun's blast of solar wind.

Ultimately, that didn't matter to them, as they knew their long journey was nearly over. They were about to be reunited with their mother. Since leaving the gravitational pull of the Earth, their wobbly paths had straightened out and become more closely aligned.

This occurred to the extent that, if their trajectories remained the same, it was more likely than not that somewhere between Mercury and the Sun the three separate travelers had the potential to join as one.

If that came to fruition, they would form an object bigger than the one that struck Earth some 66 million years ago… the same one that ended the dinosaur age.

KU1, KU2, and KU3, in theory, believed that their impact might have a cataclysmic effect on the Sun.

*****

"How many beautiful stars are shining ever so bright tonight in our Milky Way Galaxy?" Sean wondered. The night was warmer than usual on the back porch as it was partially covered and mostly protected from the wind and other elements.

Angie sat comfortably on a railing as everyone gathered close for the scheduled meeting.

After an exchange of brief greetings, she began by explaining the importance of her messages to Sean, MJ, Ellie and JP. She made it clear how she found it difficult to accept that none of the four understood the significance of Kim Alosco's article in Astronomy Magazine.

Pausing briefly for effect, Angie continued: "What place in our Solar System is filled with more known peril than any other?"

"Most certainly the Sun," JP responded.

"That is absolutely correct, my comedic friend. Do you still masquerade as a cosmologist?"

JP started to frown, but hastily turned it into a broad grin. Before he could answer, Angie parroted his response.

"The Sun!" The giant eagle flapped her wings and tapped her talons on the top of the railing.

Despite the group's laughter, Angie quickly collected herself. She preened her left wing. Between feathers, she spoke.

"There are three reasons why the Sun will be more dangerous in the coming days than at any other time in the Solar System's history. And this is why!"

Angie somberly described the travels of KU1, KU2 and KU3 since they departed the Kuiper Belt, their present position near Mercury and the reasonable probability for their conjoining.

Sheer, unadulterated shock descended on everyone's soul. Michael faced Duffy, Millie, Jack and Colleen. "At this moment, do you all have the same feelings that I have? It's as if I have been cast into the depths of Dante's damning and fiery inferno, with no hope of escape."

JP, standing nearby with Thomas, Ellie and the children had overheard Michael's words. "I assure you all, we will not descend quietly into any hell that is not of our own making. We shall fight with all our might…fight we must, never to yield!"

Intellectual denseness was not a resident among those staying in the O'Riley household. Minds feverously sorted through the bits of knowledge they had about the Sun and its menacing environs. Those minds initially thought that KU1, KU2 and KU3 had the potential to create a disaster of epic proportions.

Ellie, out of the blue, eerily recalled her phone call of a day ago with Dr. I in Hawaii during which he casually mentioned a reference to "faint comet-like tails near Mercury." That was not a coincidence.

The universe was trying to alert me to KU1, KU2 and KU3's presence. And I missed it. But I'm not sure what difference, if any, that would make now.

Ellie would not be the only one with self-doubt and regrets. Thomas also introspected. He had trouble believing, that earlier in the day, he smugly thought the world's greatest crisis had been averted. His rambling thoughts quickly reflected on the topic of a night ago concerning the 'end of days.'

He no longer wondered why they had that conversation. It was destiny! Fate, even! Some of the words to a famous "love song," of yesterday's years, streamed in front of his eyes: "Fools rush in …where angels fear to tread?"

Thomas' thoughts quickly turned despondent. *Maybe I'm the fool. Timelessness is not the critical problem at all. But all this? The Sun, KU1, KU2 and KU3 …what if there is another yet unknown catastrophe lurking? How could all this be!*

Angie had no desire to add to the spreading gloom, but she was left with little choice. Next, she repeated verbatim the words of the first message from the *voice.*

Sean and MJ, both thought that the meaning of this obtuse message was finally about to be revealed. That was not to be!

Angie looked out at the questioning faces, double punctuated with blank, conspicuous eyes.

"One of your problems in trying to interpret the *voice's* message is that most of you have become fixated on its individual words or phrases, such as 'young' and 'wisest among you.' The message can only be understood in its entirety…not in individual words or phrases. I urge you to reconsider the message."

MJ shouted, though not to anyone specific: "Why are the answers to this riddle not readily decipherable by us."

Angie started to answer MJ, but thought better of it. *My response can wait. What I have to say now, can't.*

Chapter Thirty

Winds That Speak to Trees

"Finally, I want to clear the air on one last matter. This pertains to my good friend our 'doubting Thomas' and those of like thinking. I need to inform you that your conclusions as to the existence of Mōr, the entity that Shamus met and described to you, are wholly inaccurate.

"How could you have so little faith in Shamus? With great certainty, I can reveal to you that it was Shamus, and his ally Mōr, that accomplished the feat of restoring the tidal actions in our oceans. Thanks to those two, ocean currents now flow across the entire globe assuring the continued viability of all aquatic life. I assure you, a miracle of miracles!"

It isn't a coincidence that Thomas and his fellow doubters all stand in close proximity to each other. Angie couldn't help but observe the chagrin in their faces and moved quickly to dispel it.

"For all the 'doubters' among you, you should know that Shamus does not fault you for your initial impressions where Mōr is concerned. First, you have never been participants in Shamus' previous real-life adventures. Adventures which would have provided a foundation for believing in Shamus.

"And second, this entity is exceptionally unique and exists in a realm that few of us can imagine, much less believe to be real."

Thomas, Ellie and JP, in unison, began to sing a chorus of 'mea culpa, mea culpa', thereby acknowledging responsibility for their errors in judgement where Mōr was concerned.

* * * * *

Without warning, a rush of biting wind swept through the branches of the tall, evergreen trees that served as a backdrop to where Angie stood. The tree branches violently snapped back and forth like they had just been hit by a burst of hurricane winds.

When the fierce winds first arrived, Angie quickly moved to a more protected section of the porch. The group took a collective breath when the tree branches completely stopped moving. It was as if the soft hands of an angel were holding them still. That angel was Shamus.

Slowly, the tiny face of Mōr materialized from out of the evergreens. Like a super-natural specter, a portion of its elongated body swayed back and forth behind it. Mar had reduced his size to less than *one percent* of normal, and still its body was beyond credible comprehension.

The entire group thought Angie was surprising them with a masterful illusion. Mōr's kaleidoscope of colors dazzled brightly like sparkle fireworks. It was a surreal display, yet it was awesome and beautiful.

The highlight of the illusion for Sean, MJ, JP and Ellie was the large light inside Mōr's mid-section, which danced about like a pack of fireflies. How surprised they would be when they learned that this light was Mōr's heart and soul.

Mōr grinned widely as he took center stage: "Shamus, my friend, it's so good to see you again. And of course, to greet your friends."

He turned and faced the group. "So, what do you think folks? As you can see, Angie and Shamus are not the only ones with a magical touch. And before I forget Shamus, I'm so grateful that you gifted me with the powers of a Crediach. I had a blast visiting with some of your leprechaun clans in Ireland."

Everyone had slowly retreated to the rear of the porch with an eye toward caution. Duffy and Michael looked at each other mesmerized. It struck them, at the same time, that this had to be Shamus' friend, Mōr.

Only Ellie, Sean and MJ brazenly stood their ground in front Mōr. JP and Jack, standing a few yards behind them, watched the three with a little concern and apprehension. Both ready to act if needed.

Shamus, having experienced what it felt like to see Mōr for the first time, quickly stepped forward in front of Mōr to diffuse the jolt to everyone's sensibilities.

"What a pleasant surprise it is to see you again my friend. Let me take a moment to introduce you to everyone."

All the introductions were thoughtful exchanges of kind remarks. When finished, Mōr looked out on the group with great interest. His blue eyes were full of purpose. His face seemed more like a mask, complete with a benevolent and congenial expression. It was an expression every mother would adore.

"First, I hope the timing of my uninvited appearance is not inconvenient. Before I came here, I personally reviewed all the universe's data on each of you. You are an impressive lot with a wide variety of talents and expertise.

"Forgive me, but now I can't remember which of you is the youthful and gifted astrophysicist, or which of you is the remarkably bright cosmologist."

Thomas and Duffy quickly raised their arms and pointed, respectively, at Ellie and JP. Ellie was thinking she should be terrified at the moment, standing so close to Mōr, but strangely she felt very calm. She began to think that Mōr could be an extremely interesting character. *How intriguing it would be to see into his mind!*

JP, without hesitation, walked slowly from where he was standing with Jack and joined Ellie.

"I understand that you both teach at Harvard. A fine institution, but not that golden beacon of academia and athletics that is the University of Notre Dame. I love to attend all the Irish football games. Such wonderful fanfare and spirit. And I marvel at the campus on each visit, every inch adorned with a variety of colorful flowers and tall, rangy trees."

"Ellie your grandfather attended Notre Dame, correct?"

Ellie was impressed and completely let down her guard. "Yes, Mr. Mōr, he did. And before I go any further, JP and I want to say how sorry we are for not believing you were a real entity. You and Shamus are true heroes for saving the oceans and all that lives therein!"

"Thank you, Ellie, but an apology is not necessary. I think you know that the history of science-based perceptions is replete with more mistakes…some harmless, others not…than can be counted."

"And please everyone, no formalities are necessary. Shamus baptized me with a wonderful name…Mōr…which I like very much."

Mōr now sported crystal blue eyes that brightly twinkled directly at Ellie. While he didn't understand the concept of 'love at first s

JP stepped forward with unbridled excitement. "I have always told my students that there were unimaginable and extremely powerful forces in the universe. I never dreamed I'd have the opportunity to personally meet one…an actual living being.

"Ellie and I would love to spend some time later talking with you, if it's possible, concerning dark matter, dark energy, far off galaxies and even aliens. Sorry, but right now my mind is so overloaded with questions, I'm struggling to think straight."

The reference to 'aliens' caught a sharp elbow from Ellie, smirks from Thomas, and some light laughter. The exceptions being Jack, Colleen, Duffy, Millie, Sean, MJ and Michael who were stalwart 'believers.'

A full-fledged case of déjà vu attacked Thomas. It hadn't been long since he had given a lecture in Chicago on the possibility of inconceivable forces existing in the universe.

Forces that were not only difficult for our finite minds to grasp, but such that they would warp previous scientific knowledge and corrupt, the once thought of, immutable laws of nature. *I wonder what Einstein would think if he was standing here?*

Mōr said to JP. "I'll be happy to talk to you and your friends about anything in the universe. I have many alien friends who live in the distant reaches of space. And as some of you might imagine, they have visited Earth many times in the recent millennia.

"Your early civilizations welcomed them and took advantage of their advanced intellects and technologies. But in the last century or so, they feel unwelcome here. It's a pity. Humanity could benefit greatly from their advanced intellect and abilities."

Duffy turned to Thomas. "I always told you aliens exist. Here's the proof! Why would God only create humans? At the first opportunity, I think we need to make contact with Mōr's alien friends."

Mōr paused before he spoke next.

"My friends, I'm aware that your world is presently in a timeless state. But, if I may say so, I think *you* are the ones who are actually running out of time. Angie's pronouncements are not only prophetic but they are extremely ominous."

"But before that, Jack and Colleen, would you mind if I stayed overnight? I can rest my weary body just about anywhere!"

Chapter Thirty-One

Mōr

Large bowls of buttery popcorn and make-your-own ice cream sundaes were about to reenergize the entire O'Riley household. Everyone had been, up to this point, operating on fumes.

Sean and MJ led Mōr, with MJ holding his hand, on an inspection tour of all the outside Christmas decorations and the grounds. A wide-spread grin filled Mōr's face from the sheer joy of being with the children and experiencing their human touching. Holding hands with MJ was proving to be the penultimate experience when it came to humanity.

Defying gravity, Mōr, who had no legs, merely hovered in place a few feet off the ground and was able to move his body in any intended direction through mental commands.

Mōr was amazed to see the picturesque manner in which the grounds were decorated with all the multi-colored lights and cleverly placed spotlights. Christmas time on Earth brought back other visions of wonderment that he had seen throughout the universe. *Indeed*, he thought, *Earth becomes more and more interesting every day!*

Jack and Colleen invited Mōr to consider staying in their three-bay heated garage. Even after downsizing, Mōr was still quite long. But after carefully spreading himself around the spacious garage, he found the fit perfect.

Ellie and JP were charged with the task of rounding up blankets and some pillows, though Mōr informed them that he did not need either. Having survived the freezing rigors of outer space with only the proverbial skin on his back, he was thrilled to lay anywhere in the heated garage.

Mōr willingly acknowledged to himself how much pleasure it brought him to be pampered by the O'Riley extended family. This, also, was a new emotional feeling for him and one he thought he could easily get accustomed to.

Sean, MJ, Ellie and JP had stayed with Mōr as he settled down in the garage, consuming all the popcorn Millie could make. Sean and MJ were only too happy to shuttle new bowls between the garage and the kitchen.

By now, everyone in the O'Riley household had picked out an empty spot in the the garage, where they could sit or stand to observe and listen to Mōr. The exceptions were Shamus and Angie, who sought out a quiet place to talk about events as they stood.

Mōr regaled everyone with the wonders and some of the unusual occurrences that he had witnessed in his travels through the universe. A universe that was filled with more spectacle, more beauty than one could even comprehend.

Sean and MJ sat very close to Mōr, totally mesmerized as they listened to his descriptions of stunning nebulae. "Kids, nebulae are interstellar clouds of dust and ionized gases. They are gargantuan in size and created in a million variety of shapes, infused with dazzling colors of every shade in a rainbow."

"It's so interesting to me that nebulae are one of the great paradoxes of the universe. They can form in one of two ways: either through the death of a star or the birth of one."

"Both ends of the spectrum, death and life, create a marvelous observable wonder. Far too often, people see death as ending in absolute darkness. Seldom does anyone believe that life *ends* in the spectacular light of rebirth like in our nebulae."

MJ couldn't wait to ask a question. "My dad has shown Sean and I breathtaking photos of nebulae on his computer. I think only God could create something that beautiful and magnificent. Do you believe in God?"

Mōr briefly looked at Ellie and JP, not for guidance so much as a moment to gather his thoughts. With his index finger he tapped MJ and Sean a few times in the middle of their foreheads. "What is important is not what I think, but what you believe here in your mind…and in your heart.

"I already *know* you both believe, like Shamus, that there is a God. Yet as you know, there are many like the scientist, Dr. Hawking and your own Thomas, who do not believe in a Supreme Being. What a conundrum!"

"But I can assure you all of this. There is an unseen, yet extremely notable power…a 'Presence', if you will…that can be found throughout the entire universe. A Presence that I believe is responsible for all that happens or doesn't happen. Someone or something is out there. But I have not yet come face-to-face with this puzzling essence."

Satisfied with the answer, MJ leaned back. Sean took her place and stepped to the plate. Sean said, "May I also ask you a question?"

"Of course."

"How did you get so smart and grow so big? And how do you know our friend Angie?"

Mōr responded with much of the same information he disclosed to Shamus on the night they first met. He distilled his answer not only for the benefit of Sean but also for all the others present in the garage. He briefly explained the process of how he accumulated his unparalleled intelligence by absorbing whatever knowledge existed in the ethereal spaces of the universe.

Next, he described how he eventually communicated with most of Earth's species, proudly noting that the birds, whales and dolphins were the first to respond to his telepathic conversations…which included, their friend, the witty and irresistible Angie.

"Sean, I naturally grew in size over the eons I've spent traveling across the expansive universe. The simple answer is that my body, at times, appears to grow similarly to dark matter, which instantly expands in every direction wherever it finds empty space."

"I've never eaten human food, so I can't wait to see what Millie's delicious popcorn does to me. Maybe I'll blow up!"

Sean and MJ boisterously laughed at that prospect. Ellie's ears perked up at the mention of dark matter, but she said nothing.

She and JP had decided earlier to wait until Sean and MJ went to bed before venturing into the treacherous waters of the more sophisticated and mysterious aspects of the universe.

They had no doubts that both Sean and MJ could comprehend such a conversation but Ellie had promised Colleen that she would make sure the two kept to their regular bedtime hours.

* * * * *

When the children went off to bed, except for Ellie and JP, all the adults left the garage and settled in front of the warm and glowing embers of the family room fireplace. Tea and hot drinks rested on the arms of most chairs and nearby tables. The mellowing out process was in its final stages, with soft pillows and warm beds soon to beckon all.

Shamus and Angie had finished their private conversation with each returning to the comforts of their homes. They intended to meet with Mōr very early the next morning.

Jack and Thomas came in from the cold night air of the porch, having just finished a snifter of good brandy. Jack playfully announced that a big white halo had encircled the Moon, thereby confirming his 'smell' of coming snow.

It would not be long before Colleen and Millie said good night and headed off to their respective bedrooms. Their morning prep and cooking routine would start a little earlier than normal in light of that night's revelations. They were correct in assuming the entire household planned to get an early start to the next day.

On the way up the stairs, Colleen asked Millie what she thought Mōr would like for breakfast. Millie laughed. "My dear, I guarantee you that if the popcorn is any indication, he will feast on anything we make."

After short hugs, each retreated to their rooms, though Colleen quickly looked in on Sean, and then MJ, to confirm that they were safely tucked in their warm beds. Both were fast asleep.

Thomas, now sipping a hot coffee, turned to Shamus who sat on a stool next to Michael. Duffy, by happenstance, was sitting at the far end of the couch; hands wrapped around a hot mug of tea.

"Shamus, I'm hopeful Mōr will help not only you, but also the rest of us, with the daunting tasks ahead. Each of us needs to start thinking outside the box."

"I want you all to know that contrary to my initial thoughts, KU1, KU2 and KU3 colliding with the Sun, in the routine course of events, should not be a significant threat to Earth. My recent discussions with experts in this area of science indicate that the Sun's immense mass and inherent destructive energy should be able to counter-act any significant impact.

"What really concerns me is that we need all future events involving the Sun and KU1, KU2, KU3, to come to pass under normal circumstances. There has been nothing *normal* about anything that has transpired on Earth in the last week. We must remain vigilant!"

Shamus was quick to respond. "I'm sure Mōr will do everything he can to help us. There is nothing on Earth that can compare to him intellectually, not to mention his brute strength.

"As you all know by now, Mōr has taken a fancy to all of Earth's creatures and inhabitants, particularly to Ellie. Perhaps, I should warn JP to stay on his toes, Mōr could be formidable competition."

Unabashed chuckles filled the room. When the chuckles ended, Shamus continued. "Thomas, I hope with all my heart your analysis is correct about the Sun and the three-trouble makers. But I have a gnawing feeling that it's somehow problematic. The karma is…"

Michael stood, stretched his arms and interrupted. "Pardon me, Shamus, but there is a stirring within my soul. I am reminded of an old verse I have not recited in many years which I think may be apropos for the moment."

"Unfortunately, I do not remember its author nor from where the words originated. But it goes like this:"

Kneeling on one knee, the warrior casts his long shadow
over the battlefield.
The devilish and invisible swirl disguised as 'Fate'
whispers to the Celtic warrior… seemingly spent, every fiber of his
body exhausted and writhing in pain.
'You cannot withstand the storm that is coming!'
Slowly and with great purpose, the warrior raises his head
And whispers back to Fate…
Come near, so that you can clearly hear…

I Am The Storm!!!!

Duffy's mind briefly reflected on Michael's words before he spoke: "We all need the courage, strength and mind-set of that Celtic warrior. A sprinkling of his unyielding faith and hope, would be helpful too!"

Chapter Thirty-Two

Certainty

The evening soon came to an end. Ellie and JP, after a long discussion with Mōr in the garage, stopped in the family room to say goodnight to those still up.

And Angie, who kept a watchful eye on all that was transpiring, was pleased to learn that after most had gone to bed, Duffy and Michael's promised faith discussion with Thomas was in fact had. The entire scene easily flashed before her, much like a three-dimensional hologram.

* * * * *

Michael stood; his back turned to the still glowing embers of the fireplace. Duffy and Thomas sat nearby in two large over-stuffed chairs. Shamus had left the comfort of the fireplace to join Mōr in the garage.

"Thomas, I suggested a night or so ago that we should take a moment to talk about faith. I have the feeling that the subject will become very important for most people in the coming days."

"You already know that Duffy and I believe in heaven. A world that exists beyond this one where perfect love and peace abound. We also have a deep conviction that inside each of us is a unique spirit … a light if you will…that connects *us* to each other, and to our Creator."

"I know beyond a doubt that there are many mysterious and dynamic forces that act like glue, not only to connect all living things to each other, but also to keep all that is in the universe in its proper order. Though we cannot touch or see these forces, our spirits can actually sense their presence when we maintain a high level of awareness."

Duffy, seamlessly took over. "Thomas, as you know, faith is believing that something exists even if one cannot actually see it. Like wind that blows across a beach: You cannot see the wind, but you can see the sand blow and the seaweed tumble about. Therefore, you know there is a wind… it exists."

A thoughtful Thomas stood and waived his arms in peaceful surrender. His darting eyes alternated back and forth between those of Duffy and Michael.

"Bravo, for your insightful dissertation. My heart dances with delight! As for my mind…there will always be an undefinable chasm, a grave doubt, where religions and their faiths are concerned."

"I have always believed that one must act each day with gentleness toward every living thing, and forgive everyone without exception, even when they may hurt you. I, for one, am a firm proponent that the time-honored and enduring axiom…the spirit of Christmas… should be one of life's *daily* tenets of conduct."

"Peace [on Earth] and goodwill to all."

"When will the world's religions, 'holier than thou' as they are, encourage the living of one's life by this simple axiom? So many diffuse religions…so many competing deities. Most of them, to me, are driven by greed and self-preservation at all costs. Many are reputed to be the wealthiest institutions in the world, despite professing to care for the poor as Christ, Muhammad and Mahatma Gandhi did."

"Thomas," Duffy intervened, "it is well known that these wealthy religious institutions knowingly horde their millions in gold and do nothing constructive to lighten the burdens and sufferings of the poor. For far too long, the same religions have treated woman as second-class citizens. Their guilt-ridden dogma is merely a farce to control the masses."

Thomas charged. "Then Duffy, that being said, why do so many blame an invisible God for permitting these travesties and so many others to exist? The *real* culprits are the men that govern, dominate and dictate the spheres of influence in our century's old religions; religions that allow the rich and powerful to trample on and exploit the poor.

"At least in our government, a republic, we have the opportunity to vote out the bad actors. But churches insulate themselves and perpetuate their self-interest, remaining immune from the harsh realities and objections of those they were purposed to serve."

"Thomas, I appreciate your point of view," Michael said. "I have often wondered why God does not thrash and bring such religions to their knees. Churches that are so bent on making a mockery of His stated desire for love, kindness, respect and equity for all mankind. I think it's time that we had more zealots about to remove the 'money changers' from the temple?"

"Maybe, I'm that old zealot," Thomas responded. "And I'll bet no one has ever thought to ask you two this question. Is it unreasonable to ask where your God's fanciful legions of angels have been?"

"Angels that are chronicled with possession of immense powers, angels that are capable of lifting up the downtrodden, the poor, and the disadvantaged. Why is their most celebrated appearance, in our world, only to shepherds in a nearly deserted field some 2,000 years ago?"

"Duffy and Michael, I'm more than tired, so I will finish with this. Since time began, your God, by theological definition, is well intentioned and inherently virtuous. But humanity is not! How could a virtuous deity be responsible for such a questionable creation? And you wonder why I have trouble believing in God?"

"Thomas, in many respects, I find myself in complete agreement with you," Duffy responded.

Michael silently nodded his assent. He looked at his two late night companions. "I yield! The three of us are exhausted and greatly in need of sleep. Let's go!"

* * * * *

Angie's hologram collapsed into misty space as the O'Riley household, including Mōr, began to drift into a deep sleep. How could Angie and her friends know that the darkness of this night was not about to end. Like a never-ending nightmare or a spreading malignancy, it would fill tomorrow's dawn and many of the days to come.

Part Three

Chapter Thirty-Three

Karma

December 22

As Jack had predicted, the snow fell heavily all night and well into the dawn. The Boston area would be completely shut down at least another day or two. Schools would remain closed for some time. And the White Christmas effect, as some called it, might last until spring arrived to melt the snow.

It stopped snowing immediately after the first light of dawn. Shortly thereafter, Jack found himself flabbergasted. He walked around the perimeter of his home, snow shovel over his shoulder, checking the driveway, all the sidewalks and even the approaches outside his property.

All were clear, courtesy of a gargantuan, cosmic entity from beyond the Solar System. An entity that was presently living in his garage. With a thankful grin, Jack shook his head and returned the snow shovel to the storage closet attached to the side of the garage.

Mōr, making himself invisible, had used his monstrous strength to remove the snow. While doing so, he acted and looked much like a long-tailed fire-breathing dragon, without the fire. Using his strong tail to sweep away all the deep snow and his gale-force-like breath to blow away whatever remained.

Thankfully, no one from the public was around to witness the spectacle, which would have certainly caused quite a stir in the village.

The second surprise of the morning would produce a numbing and prolonged shock. Outside, the bright sun glistening off the snow was blinding. But inside the O'Riley home, a spreading darkness of dread began to seep its way into the minds of all that heard the terrible news from Hawaii. That dread would become all-consuming in the days ahead.

* * * * *

It had started the previous night, when Ellie's cell phone rang, waking her from a comfortable dream. The surprising time of the call was not lost on her. *Who could be calling at this late hour?* She answered the call, and without speaking immediately recognized the panicked voice of her Hawaiian friend Dan Isoki.

"Sorry to call so late, Ellie. This afternoon I was air lifted by helicopter from Oahu back to the Institute. My associates had frantically called me with what they deemed to be a priority red-alert emergency. Remember in our last call I referred to "faint comet-like tails near Mercury."

"Yes, I do."

"Well somewhere between Mercury and the Sun those comet-like tails made an abrupt 180-degree turn and headed back toward the inner planets of our Solar System. What struck me as *extremely* unusual was their rate of speed. It's a velocity so fast, that we can't even figure out how to accurately calculate it."

"What makes me even more alarmed is that, assuming the objects do not deviate from their present trajectory, they are headed *directly* for Earth. And these are abnormally large objects. Not one, not two, but three! We hope to be able to determinate their sizes by early tomorrow."

"Also, you should know that my intuition tells me, with some certainty, that these three objects may be our missing KU1, KU2 and KU3.

"Dan. Worst case scenario, when will they arrive near Earth."

"Ellie, I don't think they will merely arrive. Our earliest projection is that they will *strike* the Earth sometime around Christmas Day."

For a moment, Ellie couldn't respond. She numbly stared into the vacant spaces of her room. A moment later she screamed to herself: *"Oh, my God! That would create an event similar to the one that resulted in the destruction of the dinosaurs close to 66 billion years ago."*

Anguish filled her consciousness.

"After a difficult discussion, we have decided only to alert the world's leading scientist and NASA. We also think it would be wise to hold off, as long as we can, with releasing this info to the news media to avoid panic in the general populace.

"I will keep you updated on all new developments. Let's do a video call later this morning."

Ellie got out of bed, put a robe on and went to JP's room. She knocked on the door. JP was not the best of early morning risers, but when informed of the subject of Dan's call he jolted wide awake.

"We need to see Thomas immediately," he said.

Ellie and JP softly knocked on Thomas' door which was quickly opened. He was notorious for being a light sleeper. After hearing Ellie's news, Thomas' one-sided response was short and to the point.

"That's extraordinarily fast. Christmas Day? I don't think we need to wake anyone else up right now. I'm going to call a few of my colleagues to advise them of the disturbing news. Now you two, get to bed and keep the faith! I doubt we will have the opportunity for much sleep in the next few days."

Needless to say, sleep the remainder of the night would be elusive for the three.

* * * * *

KU1, KU2 and KU3 had sped past Mercury and began their final approach to the Sun. As they did, from an unknown location within the Sun, a bellowing and rumbling wave, spouting long electrically charged tentacles of immense energy, spoke to them.

"Where do you think you three are going and why are you here? I warn you that your total obliteration is nearer than you realize."

KU1, KU2 and KU3 slowed their speed and responded to the unseen speaker with their history and the desire to be reunited with their mother, the Sun.

"You are as dumb as the rocks you are composed of and utterly misguided. The Sun is not your mother, your mother is the planet Earth. And I'm sure, she wants no part of you just like the Sun. You belong in the Kuiper Belt. Go back there now, before I destroy you all!"

KU1, KU2 and KU3 promptly turned around. Blazingly, they had no intention of returning to the Kuiper Belt.

* * * * *

Angie had received a message of the startling news before waking, which she conveyed to Shamus and Mōr when they met after daybreak in the garage. When Shamus learned the horrific news about KU1, KU2 and KU3, his thoughts drifted back to last night's conversation with Thomas. *If the karma was not good then, it's outright nightmarish now!*

The O'Riley household was scattered about, in shock, each not yet knowing what to say or do. Thankfully, the early news programs hadn't yet mentioned anything about KU1, KU2 and KU3's eminent arrival.

Thomas was sitting with Duffy and Michael in a sun-warmed and secluded corner of the porch.

"This morning's disastrous news from Hawaii gives me more than just pause," Duffy gloomily reported to his two companions. "I shudder with a level of apprehension I have never experienced. It's so difficult to consider the sheer magnitude of the terror and horror that may soon visit humanity."

Thomas replied with tongue in cheek. "Perhaps we should all sit down and write our obituaries. If the worst should happen, there will be no one left to read them. But at the very least, before doom strikes, we can read them to each other … and humor ourselves."

With a spontaneous belly laugh, he went on: "I have great *faith* that we can survive this unwelcomed danger. I just don't know how yet. In the meantime, spread the *faith* that 'we shall overcome'! Remember lads, as Yogi Berra often quipped: It ain't over till it's over."

Michael was not certain if Thomas had used the word *faith* in a serious vain, in mockery or simply in jest.

Chapter Thirty-Four

Earth

After breakfast, Colleen and Millie decided to go to the supermarket to stock up on food, beverages and other necessities with the secret goal of finishing some last-minute Christmas shopping.

They both agreed that continuing their normal routine as best they could, despite the stupefying news, would serve them and everyone at home the best.

* * * * *

Mōr, at last relaxing alone in the garage, suddenly realized that the important date of January 4 was fast approaching. This was when the Earth would reach perihelion, its closest point to the Sun during its annual orbit around the massive and fiery star. A picture of Earth's orbit flashed in his mind. Its orbit was not a perfect circle, it was elliptical… a slightly oval shape like most eggs.

Obviously, the significance of this was not lost on Mōr. It meant there was one point in the orbit where Earth is closest to the Sun, and another where Earth is farthest from the Sun. The closest point to the Sun occurs in early January and the farthest point in early July.

The fact that the Earth would soon reach perihelion, meant that KU1, KU2 and KU3 arrival at their intended destination would take less time and distance than first thought. *I think* it's *important that I provide this data to Jack and Thomas' think-tank as soon as possible.*

* * * * *

A command post had been set up in Jack's spacious study off the family room. Double computer screens were banked in such a fashion that they could be viewed from anywhere in the room or its entrance. Multiple surround sound speakers in the study added an eerie theatre touch.

Jack, Sean and MJ had joined Ellie, JP and Thomas in front of the computers, eager to participate in the early morning video call to Hawaii. Duffy and Michael had ventured over to the garage to discuss matters, as they stood, with Shamus, Angie and Mōr.

When the call began, the first thing Dan expressed was how happy he was to actually see MJ and meet the other participants as real-life people. Then it was down to business.

First, Dan reported that it had been impossible to keep the press and other media sources from the catastrophic news about the approach of KU1, KU2 and KU3. Leaks were springing up everywhere.

Second, he stated that the powers in Washington, D.C. had advised NASA to prepare a draft speech for a presidential address to the nation later that night.

Dan then considered the most obvious questions on everyone's minds.

"There has been much discussion of altering KU1, KU2 and KU3's projected course via spacecraft. But of all the world's known space agencies, none are prepared or equipped to launch a manned space vehicle, or even an unmanned rocket, into space within the next two days."

"And the International Space Station, despite its low orbit around the Earth, also lacks any useful capabilities to thwart the threat. By international agreement, the ISS is intended to be shared and used by its member nations exclusively for peaceful means by providing a space environment for scientific research. The development or the placement of any offensive weaponry thereon has been strictly prohibited."

Over the next hour, or so, animated and passionate discussions flowing back and forth, contemplated most of the conceivably logical and theoretical options one thought possible to prevent the looming diaster.

And that was the problem, though all the reasoning, both theoretical and scientific was sound on its face, it produced no viable solutions for the near term. Practicality, like a phantom, was elusive!

The overall outlook appeared bleak. Perhaps even hopeless! Jack and Thomas, like everyone else in the room shared the same opinion. Simply put: Earth did not possess the necessary knowledge and tools to prevent KU1, KU2 and KU3 from destroying it.

With mental fatigue setting in, a break seemed necessary. A joint video call was planned for later that evening.

Thomas immediately left for the quietude of the outside porch. After a period of introspection, he sought out some of his fellow associates throughout the world. Each agreed to update and recheck all their data, specifically the present whereabouts of the "invaders", as they were now called for simplicity's sake.

* * * * *

JP was feeling rambunctious. With a devious smile, he shouted. "I bet Sean and I can make a snowman that is more exquisite than anything Ellie and MJ could possibly conjure up."

"Your either delusional or you've lost your marbles, JP," MJ shouted back. "Which is it?"

JP's wily challenge was accepted and all four rushed to put on suitable gear for the contest. The lack of wind, combined with a Sun that warmed all that it touched provided the perfect conditions for snowman building.

Thomas, meanwhile, had grabbed another cup of coffee and decided to head for the garage to bring everyone there up to date on the latest developments and brainstorming. He was surprised to see that Mōr, Angie and Shamus were nowhere to be found. The three had disappeared without leaving an explanation.

Thomas went outside the garage upon hearing the shouts from the snowman builders. He spied Duffy and Michael walking down the long hill of the driveway, apparently on an exercise mission. He found Jack nearby, loudly cheering for MJ and Ellie's team.

Jack informed Thomas that Angie had kept Shamus and Mōr, up to date on the somewhat outlandish and tenuous ideas that were floated during the video call earlier that day. Thomas didn't need to question Jack as to how Angie had known what was discussed.

* * * * *

Angie had laughed to herself when she observed the spirited contest between the snowman builders. A little unwinding was healthy for everyone. Then she took flight, knowing the seeds of thoughtful inspiration had been well rooted.

She now sat high on a thick evergreen branch that hung next to the front door of her home in the immense forest of St. Johnsbury. She wondered how her allies were doing? Early that morning when they convened in the garage, a plan was hatched.

Shamus would go to the North Pole to make Santa aware of recent events, which could gravely affect his annual Christmas Eve sleigh ride.

Meanwhile, Mōr would travel as fast as he possibly could to the farthest corners of the universe in search of help.

* * * * *

Sean slipped and fell in front of JP, just as they placed the featureless head of the snowman on top of its body.

"Sean, give me your hand. You don't want to give the girls any hope by them seeing you laying down on the job."

JP quickly pulled Sean to his feet. When fully standing he whispered to JP.

"Do you know who the Avengers are? Laying on the ground, I was just thinking how much we could use their help now."

"I'm a big fan," JP smiled.

Sean, in a hushed but excited tone continued. "My favorite character is the courageous and dashing Captain America with his indestructible vibranium shield."

"I agree with you that Captain America is fantastic. My favorite is the charismatic Thor, such a heart-warming character. He's so powerful because his energizing four-square hammer, named Mjolnir, is made of magical uru metal."

"Sean, did you know that one of Mjolnir's greatest powers is that it has the ability to create force fields, capable of initiating gigantic explosions mighty enough to destroy a galaxy. We could certainly use that…"

An unseen shift of mental energy, perhaps better described as an insight, emerged from one side of JP's brain. And then a word flashed several times in front of his eyes, like the action of a broken old-fashioned movie tape. *Shield… shield…shield…shield!*

The movie tape came to a stop: *What shield…why a shield? Maybe…*

JP pivoted and ran to Ellie, nearly screaming. "Time out! I think Earth needs a shield."

"JP, the Earth already has a shield. It's called a geomagnetic field, commonly known as Earth's magnetic shield. It protects us from the harmful effects of the Sun's hot blast of solar wind and ultraviolet radiation."

"I know that, but right now, I can't remember the nuances of how it works, Can you quickly explain it?"

"Of course, silly. The magnetic field is spawn by electric currents due to the motion of convection currents of molten iron and nickel in the Earth's outer core. It extends some 10,000 kilometers, a little over 6200 miles, into space.

I'll use the cookie coloring, we brought out to detail our snowman, and draw a diagram in the snow of what the magnetic fields look like."

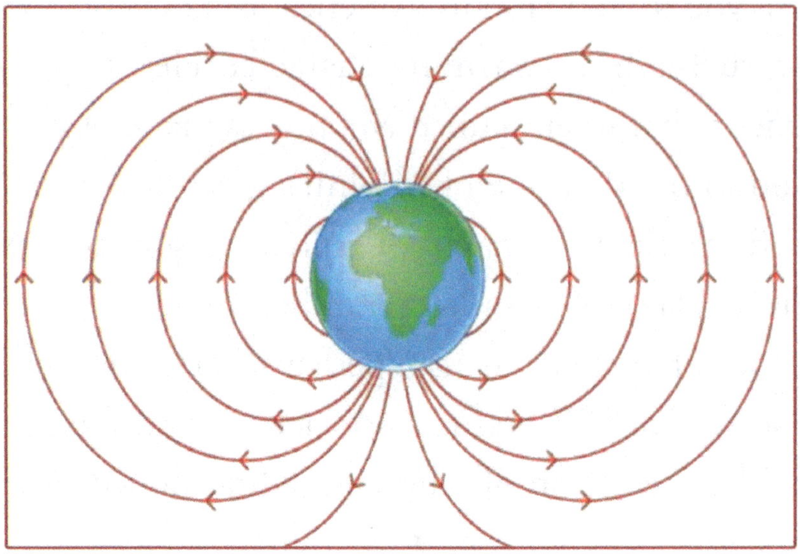

Earth's Magnetic Field

"Ellie, we need Thor's hammer to super energize and electrify the magnetic fields, to an unimaginable power equal to at least five hydrogen bombs. Then we might be able to stop the invaders."

"JP, I trust you understand, Thor does not exist in our world."

"So, what! We need to create something similar to his Mjolnir! I declare 'no winners' in our snowman contest. We need to get our group's think-tank together ASAP."

JP's shouts to Duffy, Michael, Thomas and Jack could be heard all the way back to an enchanted forest in St. Johnsbury.

Chapter Thirty-Five

Friends Indeed

Mōr had moved in excess of the speed of light through the Milky Way. He checked his data base to confirm that he had arrived at the precise location needed, slowed and came to a stop. Floating in space, he initiated the sending of coded signals to planets whose homes lay within the galaxy of his birthplace: EGS-2s8-1.

These signals, if they could be seen, resembled continuous rolling wavelengths, moving multiple times *faster* than the speed of light in a vacuum. Like Mōr, these wavelengths were unlike any known to the physicists of Earth.

The signals arrived at their destination, near the Kuiper Belt, shortly after being sent. Mōr was grateful. Sending the signals saved Mōr a tremendous amount of travel time, that would have been required if he had to personally deliver his urgent and desperate plea.

Time. Mōr was absolutely positive that he was not responsible for Earth's loss of time. He had long been of the opinion that time, at least on Earth, had moved to or been gobbled up by a parallel dimension. He had made no determination if this was temporary or permanent? Weird to be sure!

Thankfully, out here in the universe, the concept of time continued to exist. Given the deadly threat approaching Earth… it was more vital than ever before. Not a second could be wasted.

Good fortune would smile on Mōr. Within moments of each other, he received two responses to his signals. *Shamus must have attached some lucky green shamrocks somewhere on my person.*

Mōr's far-away friends indicated they were more than happy to assist him, and would depart ASAP, after making quick preparations for the journey.

It was agreed that the best place to meet was near the constellation known as Orion the Hunter, the Northern hemisphere's brightest winter constellation. More precisely at Sirius.

Mōr was very familiar with Sirius, because it was the brightest *star* in Earth's night sky with only the Moon, Venus, Mars and Jupiter appearing brighter. And it was close to Earth, only 8.7 light-years away.

The Force's friends were from the planets Zillox and Tradown, planets located within two light-years of each other in a galaxy twice the size of the Milky Way.

Despite the distance between the planets, the denizens of Zillox and Tradown were very friendly neighbors. More interesting, was the fact they looked exactly alike. Facial variations, if any, were minor. The odds of such an incredible similarity in appearance had to border on the absurd.

The inhabitants of Zillox and Tradown were among the most advanced, both physically and intellectually, of any species in the universe. Quite remarkably for both, their methods of communication were entirely telepathic.

Each civilization had conquered the highly complex nuances of time travel. They both possessed the ability to transport their space vessels across the spectrum of the universe nearly instantaneously. To them, warp speed, was obsolete.

Humans would have called the inhabitants of Zillox and Tradown "aliens", but to Mōr they were simply old friends. They planned to arrive at the agreed upon location in three vessels, all identical in shape and size… within seventeen minutes.

The main deck of each vessel was more than thirty times the length, and at least twenty-five times the width, of a standard NFL football field.

To give some perspective to the immensity of the vessels, one had to conjure up a scene from Episode IV: A New Hope, of the Star Wars saga…the biggest and most powerful imperial warship ever created…Darth Vader's Super Star Destroyer. But the difference between the alien vessels and the imperial warship was just that, the vessels were not intended to be warships or hunters.

Each vessel had 80 decks, with the 35 levels above the super- structure fronted, from floor to ceiling, with specially developed clear glass-like panels. This enabled panoramic viewing into space from any position in the front portion of a deck. Novel to each vessel was the fact that the top two decks served as its command and operations center.

The remaining 45 decks were located below the super-structure's main deck. These contained the power plants that provided the energy for every aspect of the vessel's operation, including the mechanicals for day-to-day operations. Well-appointed and generous officer and crew quarters, gourmet kitchens and spacious recreational areas were located on decks 40 to 45.

The recreational decks included movie theaters, fitness training areas, yoga rooms, an Olympic size heated swimming pool and well-staffed spas.

On all the command decks, individual groups of Mōr's friends now stood with both arms crossed over their chest and heart. In their galaxy, this was the universal sign of love and peace.

With a waggle of his tail, Mōr acknowledged their kind gesture, and the light within his body momentarily expanded and glowed like a bursting star.

She stood in the forefront of the 80th deck. Statuesque, with long braided strands of blonde hair that fell neatly down her back and past her waist. Her blonde hair complimented and enhanced the glow of her smooth mahogany-colored skin.

In stark contrast to humans, she had no ears, a facial trait shared by all the members of the armada. And yet, some would find it a strange coincidence, that liquids, nutrients and foods were processed through a mouth placed and shaped similarly to humans.

Her light-pink lips, lent an air of gentleness to her otherwise sizable oval face. Her unusually large head provided space for her enormous brain, which was more evolved and intellectual than any other in the universe.

To the uninitiated, it appeared that her striking green, egg-shaped, eyes stared blankly off into space. In fact, they were zeroed in on the beautiful Orion Nebula in the constellation of Orion the Hunter.

Stupendous, widespread and polychromatic, the nebula had a mass 2000 times that of the Sun nearest to Earth. Known as one of the brightest of all nebulae, it could be seen from Earth and identified with the naked eye as the middle "star" in Orion's sword.

Not immune to the plight of most females, she took a moment to adjust her form-fitting, uniquely designed tunic. The tunic, woven from elimia, a silk-like fabric, was resplendent in bright colors and greatly flattered her six-foot, two-inch slender frame.

A bright, gold shoulder sash crossed her chest, and then looped around and was tied off at one side of her waist. Its two long ends dangled freely down the side of her left hip. The sash an insignia of ultimate command.

She was the admired, and much loved, senior admiral of this allied armada. Without a doubt, she would have been thought of as exceptionally and strikingly beautiful in any galaxy.

Shanta raised one of her long spiny arms and ordered the three vessels to cautiously approach closer to Mōr's location.

Urgency dictated that customary greetings and catch-up conversations be dispensed with to a later time. Mōr's initial plea had summarized the critical status of the Earthlings and their tragic inability to prevent a direct strike by KU1, KU2 and KU3.

Shanta informed Mōr that the senior officers of all three vessels, had established a fleet task force, which was already monitoring the actual movement and location of KU1, KU2 and KU3 in relation to Earth's orbit.

The task force, composed of the brightest experts and various specialists in the armada, had been analyzing calculations and possibilities during their journey through the Milky Way. It was in this crucible that initial possibilities would give way to probabilities, which could then be whittled down to reason-based projections.

When Shanta communicated with Mōr, she had the benefit of viewing floor to ceiling, and wall to wall, futuristic colored screens. The screens were capable of being split into smaller and distinct sections to depict a multitude of disparate events. Three such sections now displayed the top three projected scenarios of KU1, KU2 and KU3's future activity.

Each projection was delineated on the screens in a different color and superimposed over each other. Young children would have found the systematic construction of colorful lines whimsical in appearance. It was anything but!

"Mōr, the computer models on my screens have forecasted that KU1, KU2 and KU3, which Earthlings now refers to as the 'invaders', will strike the Earth in the Pacific Ocean. More specifically, it will be a direct hit on the Hawaiian Islands."

"Shanta," Mōr asked: "Have you been able to determine the size of each invader, their present velocity, and their predicted velocity at impact?"

"Yes, my friend, we have. Our projections do not allow for any margin of error. *Each* invader is roughly six kilometers, or 3.728 miles, in diameter. More importantly, it really doesn't matter much if the invaders conjoin or strike in close proximity to each other. The result will be the same…total destruction of the planet."

Shanta would continue her monologue dispassionately, with truth's hard facts that would leave any listener numbed.

"The invaders will strike with an aggregate diameter of approximately 18 kilometers or 11.185 miles. While the three objects have traveled through the universe at unnatural velocities, we believe the Earth's atmosphere and other factors will cause them to slow before they actually strike. Impact would occur at close to 70,000 miles per hour."

"Shanta, Earth was struck by a meteor or an asteroid some 66 million years ago. That object was thought to be six miles in diameter and struck the Earth at roughly 40,000 miles per hour, terminating all dinosaur life. These invaders are close to double that size and will arrive at twice that velocity. So, are my internal projections correct when they indicate that…"

"Yes," Shanta, interrupted. "*All life* on Earth will be completely eradicated. There is no hope for survival or possible regeneration at any level."

"Our models project that immediately after impact with the Earth's crust, ocean waves will reach heights that will be five to eight miles high, at minimum. They will surge relentlessly against every continent. Their destructive powers on reaching land masses need not be elaborated."

"In addition, Mōr, the impact will form a crater ten times the size the state of Alaska. Super-heated chunks of Earth's crust will be blasted out into the upper reaches of the stratosphere."

"Gravity will pull most of that fiery rock and debris back down to Earth, heating the atmosphere to temperatures never thought possible."

"In less than a day, the extreme heat fanned by ferocious winds, far in excess of any known hurricane, will ignite firestorms around the Earth. This endless burning firestorm, in turn, will create a thick dust and debris filled covering over the entire globe, thereby plunging Earth into total darkness.

"The result will be that for untold years, this covering will completely block the light of the Sun. And as we all know, natural food supplies, without sunlight, will quickly vanish."

"Adding to this misery and destruction will be global volcanic and earthquake activity, on a magnitude never before experienced. The landscape of Earth will be drastically altered to the extent that it will not be recognizable from its present form."

"Sadly, Mōr, these cataclysmic convulsions will last for many years. But eventually they will slowly moderate and cease."

"Every species of animals, aquatic life and all of humanity will eventually perish. Anything that survives will fade in less than a year or so, due to the total annihilation of the environment. Lack of food and oxygen deprivation being the primary culprits."

"The Earth will become a place of extreme heat trapped underneath the global dark covering. The trapped heat producing a planet-wide landscape that is nothing more than a barren wasteland… just like that of Mars."

Mōr cringed. "Shanta, have you been able to formulate any plans or ideas that might give Earth hope?"

Chapter Thirty-Six

Agony and Tears

December 23

The morning dawned with every church, synagogue, temple and place of worship, prayer and meditation bursting at the seams. And yet, others simply stood outside under a canopy of stars, or the warmth of the Sun, or intemperate skies and lifted their eyes and voices toward the heavens.

Their prayers asked that this cup, filled to the brim with the bitter taste of agony, pass from their lips.

Agony presented itself in many forms: pain, torture, distress, torment, misery, grief, trauma, sorrow or heartache. At this very moment, there was not one inhabitant of Earth who was not experiencing at least one, if not a combination, of these symptoms.

The believers in a loving and compassionate God probably constituted a majority of the world's people. It would not come as a surprise that incomprehensible fear had driven the non-believers, at least for the time being, to feverishly join with the believers. No conversion rituals deemed necessary.

Paralyzing fear was a curious emotion. It was capable of causing the weak to withdraw and hide among the disquieting shadows cast by that which was feared. But there would always be those who could find courage, those who could instill themselves with bravery and faith to confidently, and even aggressively, overcome their fear without trepidation.

The future was about to redefine the look of courage.

* * * * *

Colleen and Millie had performed their usual culinary feat, cooking a hearty breakfast of fresh-baked muffins, eggs to order, bacon and ham. The harmonious noises of a satisfying breakfast seemed to bound everywhere through the kitchen area.

During a serious moment the O'Riley clan openly vowed to banish panic and fear from every thought. Minus Thomas, most would take comfort in their belief in a loving and caring God. In the ensuing hours, silent prayers would endlessly ascend from their minds and hearts. Prayers instinctively mumbled by passive lips.

Given the reality of the world's circumstances, the household represented a small microcosm of the world's population. Anxiety simmered just below the surface of everyone's masked attempt at calm and normalcy.

The entire array of life's emotions was on display. Collectively and individually. Some obvious, most hidden. Feelings *no one* could avoid. Understanding the stark realization that this appeared to be one of the last normal days on Earth…how would you expect anyone to feel?

It would be Thomas' confident voice that temporarily halted the forks and knives dicing up breakfast. "Remember my friends, God only helps those that help themselves. Now is the time, if there ever was one, for all of us to band together and keep unshakeable faith in each other. By doing so, we might find a way to avert the impending disaster."

"I encourage you all to pray as hard as you can. But in the meantime, it is vital that we push every ounce of our imagination and intelligence beyond their limits. Life on this planet depends on it."

Jack winked at Duffy and Michael with one thought. *What a metamorphosis Thomas just had, even if it's only temporary. He gets it!* Jack instinctively pumped his right arm high, extremely proud of Thomas and his pronouncements.

Few noticed that a somber Thomas, now sat pensively.

I know my words contradict nearly everything I espouse. But my ship is not sailing under ordinary circumstances. The seas are ominous and treacherous. The waters extremely turbulent and frenzied, not calm. The onslaught of high menacing waves unusually vicious.

If I - meaning we - at this very moment do not support and help each other in any way we can, that ship is going down with all hands-on deck…and dragging with it many more unfortunate souls!

No one loitered after breakfast, so a happy Jack suggested a long stroll to relax a bit in the sunny morning. He casually added that Angie had earlier informed Millie that Mōr was expected to provide a status report in about an hour.

* * * * *

The walk turned out to be a perfect elixir. The chilly air was crisp and refreshing. Warm rays of sunshine danced and reflected off the snow and ice-encrusted trees.

If Angie were present, the scene would remind her of sunrise at home in her northern forest. This morning's viewing was truly astonishing. It resembled a crystal cathedral filled with a kaleidoscope of dazzling colors.

Colleen held tightly to Jack's arm as they slowly walked, while Millie mirrored the same position with her much-adored Duffy. Side by side, the old friends talked about nothing and laughed about everything.

Millie fell silent for a moment before speaking. "Life on Earth can be so incredibly beautiful. Look at this crystalized rainbow of a morning! It reminds me of the colorful landscapes you read about in fairy tales."

A soft chorus of voices approved her keen sentiments.

Ten yards behind, Ellie's arm was firmly entwined with Thomas' own while her free hand clutched JP's. She caught a glimpse of growing mist in Thomas' eyes. The mind behind those eyes filled with one thought. *This may be my last opportunity!*

"Ellie, I want you to know and hear from my own lips…that there is no one that I love and admire more than you."

Her grasp grew tighter and her head gently drifted to Thomas' shoulder. "Thomas, you're so kind. I will treasure you and your words all my days, despite your irreverent nature."

Thomas' inner being glowed. *This for me is, by far, the most fulfilling of all feelings.* He turned his head toward JP.

"JP, you've got a heart as big as the Atlantic Ocean, coupled with a gifted mind and witty, charming ways. Our world is in desperate need of your talents and I trust I'll have the opportunity to witness your many successes. Something tells me you and Ellie, each, will have a wonderful life."

Now is not the appropriate time to tell Ellie and JP that I really meant to say 'share' a beautiful life instead of 'have.' I'll remind them someday that love has reasons of which reason knows nothing. And that, when and if they do love, they should love each other blindly to the exclusion of all life's vagaries.

JP warmly patted Thomas on his back. "I'm so grateful that Ellie brought us together."

In a few more steps, Ellie became lost in thought in a far distant world. She vividly recalled JP's words on the subject of fear after skiing and pizza. Once more a strong realization fluttered her way, affirming that JP was so much more than just her new best friend. Warm fuzzy feelings wrapped themselves around her.

What a shame this stroll can't last forever, Ellie thought. *We are not without hope and everyone has a realistic understanding of what the future may hold. In that understanding we are finding and sharing beauty.*

* * * * *

Michael had decided to remain behind and entertain the late arrivals for breakfast. Sean and MJ, seated at the kitchen counter, were drinking mugs of hot chocolate.

In the middle of discussing the latest Celtic basketball failures, Sean looked up at Michael. "I think we have to do something very special in an effort to help solve our present situation."

"Sean, MJ. I'm not sure I told you this before, but to do or be a part of something special, you have to find something special *inside* yourself first. We all believe both of you are very special. I hope *you* believe that too. And if you don't, you should!"

Sean and MJ looked at each other and then back at Michael, nodding their heads.

Michael stood to clean the kitchen counter. He found his mind wondering as he did so: *Saving the planet will surely take something or someone extraordinarily special. Problem is, no one at the moment knows what…or who…that might conceivably be.*

Chapter Thirty-Seven

First Probe

Shamus had just returned from the North Pole, where preparations for tomorrow night's eventful sleigh ride were nearly complete. It was mid-morning, when he met Angie alone in a warm corner of the O'Riley back porch.

Mōr had just finished relaying to Angie all the particulars that transpired between him, the Zilloxes and the Tradowns. Angie now updated Shamus on that status and thereafter would do the same with her friend Millie.

* * * * *

Shanta, agreeing with the advice of her task force, had decided to move the armada from Orion to the outer fringes of the Earth's magnetic field in the Western Hemisphere. They now lingered above the northern portion of the Pacific Ocean.

The task force estimated that the KU1, KU2 and KU3 were less than one Earth-day away. Mōr had suggested that it might be beneficial to open up a line of communication with his friends on Earth, to keep them informed of future developments as they occurred.

Shanta was reluctant to do so at first. She requested time to first review the backgrounds and credentials of those at the O'Riley homestead. The review did not take her long.

"Mōr. That's an impressive array of intellects and experience you have gathered in that Boston home. I'm impressed. Let's give them a chance to give us their input and provide them with real-time visuals of developments as they occur up here."

"I'll have Captain Zken arrange the process on our end. Please alert the Earthlings of our intentions. I would also like to bring the talented and renowned duo of Angie and Shamus into the loop. Their history, in difficult situations, is extremely remarkable and should not be ignored."

"I know time has ceased to exist on Earth, so when the visuals are working, I'll arrange for some necessary context to our communications and interaction with the O'Riley household."

"I'm going to order that the armada adopt Earth's year, month and day calendar and construct a time clock based on Boston's normal eastern standard time schedule. That should provide some needed order for everyone."

*****

The first images received by the O'Riley computer screens were streamed from a small exploratory probe sent out by Shanta on the armada's arrival near the magnetic fields. The probe was solely programmed to monitor KU1, KU2 and KU3's as they neared Earth.

It would shadow the trio of invaders, hopefully undetected, some 150 kilometers, or a little over 93.21 miles, from slightly behind and above them.

The overflowing O'Riley command center was jammed with every member of the household. Its center and entrance bulged with bodies jockeying for position. All froze when they saw Shanta. The look was one of total disbelief.

Those in the O'Riley household were not the first Earthlings to actually see a living alien. Nevertheless, as each screen displayed images of the aliens moving attentively in their various work environments on the vessels, all fell silent.

Mōr's reveal, all by itself, had been breathtaking. This…this was something else!

Sean was the first to point out that the aliens were all tall and slender. MJ observed that the aliens were actually quite graceful and fluid in their movements. She thought them beautiful in appearance. She whispered to Duffy, "Do you think we'll ever get a chance to meet them?"

"Michael and I certainly hope so with all our heart. How foolish man has been to define himself as the only intelligent species in the universe, failing to take into consideration the existence of other far-flung more intelligent beings."

The grainy images of KU1, KU2 and KU3 approaching Earth did not appear frightening. They looked like rough, irregular boulders silently hurtling and tumbling in frictionless space, against an endless background of twinkling starlight

Innocent enough. Harmless looking. But all knew the deadly consequences. Termination.

Eyes grew wide, mouths fell agape, and arms were folded across chests. From some eye's tears fell. Those present knew all too well the unfolding catastrophe that approached their planet.

Each now a personal witness to what might very well be the last event mankind would ever call "history." A history that no one would have the opportunity to record!

Chapter Thirty-Eight

If Only

Back on the armada's command vessel, Lynkai, the youthful and energetic lead astrophysicist on the task force, stated the obvious.

"We must try to stop the invaders before they reach the Earth's magnetic shields. Because we don't know the metallurgical or molecular compositions of these characters, we can only guess as to how they will interact with the shields peculiarities."

"Personally, based on the data we have, I think KU1, KU2 and KU3 will travel through the magnetic shields rather easily. Their size and speed make them extremely formidable and powerful. I hate to say it, but its game over if they pierce the shields."

No one at the table or in the situation room, disagreed with her observations. By general consensus, it was determined that an additional highly sophisticated probe would immediately be sent out to explore the invaders' physical characteristics.

Mōr insisted that he should accompany this additional probe. Real eyes might see or sense something the inanimate probe might miss. All of Shanta's instincts told her not to allow Mōr to do this. But in the end, Mōr's cogent arguments were persuasive.

Permission was reluctantly granted, with the stern provision that Mōr remain invisible throughout the exploratory mission. An order was an order. He would obey it like any good soldier.

* * * * *

The three vessels of the armada stood only 200 kilometers or 124.28 miles apart, in a semi-circle that conformed to the outermost curved perimeter of the Earth's magnetic shield. For simplicity's sake, this perimeter would be referred to as the "outer rim." Shanta's command vessel took center stage.

Preparations for the new probe complete, it and Mōr were ordered to depart into the abyss of the mirky unknown. The first item on the probe's agenda would be an attempt at a physically exhaustive and detailed examination of the invaders. Confrontation was absolutely necessary.

* * * * *

For a while matters went just as anticipated. Mōr and the probe, thanks to Shanta's time travel, came in close proximity to the invaders almost instantly after departure. Mōr believed it was favorable, that KU1, KU2 and KU3 still remained a significant distance from Earth. There was enough wiggle room in case some unforeseen problem arose.

After taking hundreds of close-up high-resolution pictures of the invaders, from every possible angle, the probe slowly descended and landed on top of KU2. The maneuver and its execution were going as planned. Next would be the task of acquiring a sample of its composition.

As soon as the sampling arm dug a few inches into the rock, an exceptionally strong electrical current coursed through the entire probe. The shock caused the probe to instantly jump back into the safety of space. Singed, but unharmed.

No sample was obtained.

In minutes, the probe's programming was successfully reset by a now highly concerned Lynkai. A second attempt, of the same procedure, was now planned for KU3. Once again, the landing went smoothly and the sampling arm began to dig.

In seconds, another electrical surge attacked and, like the first, coursed through every inch of the probe. Unlike the first surge, this shock was 150 times more powerful than the one generated by KU2.

The probe admirably endured the surge for about 27 seconds before suddenly exploding. Its components disintegrate What little remained quickly became the latest addition of space junk floating in the universe.

Mōr and those in the armada watched the explosion in horror and disbelief. Unfathomable! Even the usually stoic Shanta was bewildered. Those observing from the O'Riley household held long breathes, only to release them with staggering dismay.

Shanta sent a transmission to Mōr.

"Is it my imagination or are my senses accurate in warning me that these objects appear to be able to *communicate* with each other. Tell me. What are the chances they know of your presence despite the fact you are invisible?"

"Unfortunately, Shanta, I think we both know the answer. It would not be a stretch to surmise that they may also be aware that your armada is nearby."

On Shanta's command vessel, the next thirty minutes would pass in swirls of frantic activity.

Chapter Thirty-Nine

Checkmate

In the end, no further consultation with the command staff would be had. The order was loudly and crisply given: "Prepare for battle stations."

Shanta was determined to put an end to KU1, KU2 and KU3. Whatever it too! Her communications with her crew would now be extremely crisp.

A scurry of bodies moved rapidly about the three vessels, all activating and readying computer-driven systems that would enable implementation of Shanta's orders.

"Power up the Riyels," Shanta directed her second in command, Captain Kictrap.

"Aye, aye, Admiral."

All the might of the armada's defensive powers was about to be unleased. Long thin cannons began to protrude from their concealed housing on the middle decks of each vessel, decks 42, 43, 44, 46, 47 and 48. Each deck held eight banks of canons.

The cannons could be rotated or moved a maximum of 25 degrees in any direction. When a vessel fired its cannons, in combination, it created a generous "field of fire" - capable of delivering *precise* streams of laser beams far more powerful than multiple hydrogen bombs.

* * * * *

The scientists of the Zillox's and the Tradown's had learned to harness the secrets of a far-off galaxy's sun, a sun that they named Riyel. Similar to Earth's Sun, Riyel, every second generated tremendous amounts of energy by a process called nuclear fusion.

Simply stated, our Sun's energy results from four hydrogen atoms undergoing a series of collisions… what remains, fuses together to form one helium atom.

Such reactions, occurring 100 million quadrillion, quadrillion times each *second*, release gargantuan amounts of energy. Just as Einstein's $E=mc^2$ predicted.

These brilliant and innovative alien scientists were than able to control and channel the raw energy of a nuclear fusion into a lethal blast of a high-density ray…a perfect laser beam.

Their insights next led them to recreate this fusion process in a specially built chamber, which they nick-named after Riyel. Such a chamber sat in the middle of the lowest floor, deep inside, each vessels hull.

The breakthrough that ultimately led to the ability to build Riyel was the discovery of a super-strong alloy of Zieken and Yojd, ores found only in the Kuiper Belt. The alloy permitted the construction of layers of casings that safely contained the volatile nuclear fusion process.

A fusion process that the scientists engineered to be *six times* more powerful than that of the Riyel sun.

Normally it took about 15 minutes to power up the chamber's raw energy to maximum power in order to produce a fully energized laser beam. A highly advanced fire control system, with its banks of screens filled with bright colorful lights and levers, would indicate when Riyel was ready to fire.

* * * * *

Shanta ordered Mōr to retreat back to the protective cover of the armada and remain invisible. He was only too happy to comply.

"Captain. Deploy and fully prepare the cannons."

"Aye, aye, Admiral. Cannons deployed and ready."

"Fire Control. Synchronize lasers for acquisition and target lock."

"Aye, Admiral."

"Captain, advise when Riyel has reached max capacity."

"Aye, aye, Admiral."

The fire control officer stared at the large rectangular screen which depicted, in various colors, KU1, KU2 and KU3 and the immediate area of outer space that surrounded them. A square box outlined in bright red framed the constantly advancing position of the three invaders. Green lines represented their projected trajectory toward Earth.

The square box borders would blink in red till a target was acquired. Distance to targets was automatically calculated and colored in navy-blue. While actual time to target scales were set in bright yellow.

The firing officer's screen was simultaneously transmitted to the command deck's, wall to wall, screens. Thus, they also appeared back on the O'Riley computer screens. All of the potent laser beams, when fired, would be directed at and concentrated within the bright red framed box. The acronym for this was "fire for effect."

Command and the task force had earlier calculated that the powerful lasers beams would pulverize KU1, KU2 and KU3 into dust or, at the very least, shatter them into harmless pieces.

The fire control officer turned from his screens.

"Admiral, Riyel has reached max capacity. Ready to fire laser beams on your command!"

Shanta repeated: "Maximum power. On my command."

"Captain Kictrap, you may fire for effect!"

"Aye, aye, Admiral"

A volley of an intense concentration of deadly laser beams from four dozen cannons… powerful enough to level the entire New York City metropolitan area, Long Island and all of New England instantly… flashed from each vessel toward the invaders as depicted on a bright red blinking square.

Hearts stopped beating in the O'Riley household. Arms and hands reached out to be held. Anticipation had reached its zenith.

Shanta, her staff, and each vessel's crew members looked on confidently.

Chapter Forty

The Devil's Dance

The firing officer's screen automatically counted down the time sequence from firing launch to impact at target. Again, this was amplified on the wall-to-wall screens and also transmitted to the O'Riley's.

"Admiral, thirty seconds to impact."

"Twenty seconds."

"Ten seconds."

"Five seconds."

"IMPACT!!"

* * * * *

The fire control officer, in the last stage of the countdown, just *assumed* impact. Only one problem. His screen, exhibiting real-time pictures from the still hovering first probe, was *blank*. Blank save for a background filled with twinkling starlight.

And the targeting area within the bright red square was completely empty. No flashes. No resulting star-burst blasts. No destruction. No devastation. No clouds of dust. Nothing. Not even KU1, KU2 and KU3!

If it wasn't such a serious matter, one would almost find the moment laughable. But funny it was not. Where could they have gone?

The command's fire control, the most technologically advanced system ever designed, registered no direct hit on the targets. That wasn't possible, practically and theoretically, given the magnitude of the fire power launched.

A review of the firing tapes was quickly ordered.

The review left no doubts. The tapes confirmed that the laser beams clearly entered and filled the area designated by the blinking red square. After doing so though, the laser beams simply passed through that area, dissipating harmlessly throughout space.

Though they had vanished, there was *no* evidence that KU1, KU2 and KU3 had been struck at all.

This disheartening realization caused innumerable questions to abound. None could be answered. The reality of the laser strike's failure was incomprehensible. Unthinkable! The seeds of chaotic reasoning began to sprout. What could possibly have gone wrong?

In the confines of the O'Riley's, all that was important in life became scrambled. Along with a tsunami of sobering doubts about Earth's ability to survive. Doubts that would soon turn into a paralyzing fear.

Now it became clear, an inevitable reality no longer escapable. The future would be filled with unadulterated terror.

* * * * *

KU1, KU2 and KU3's ability to evade the laser strike, and subsequent disappearance, had never once been considered by the armada's task force in the equation of possible outcomes. Shanta stood, by herself, staring off at distant stars in the infinity of space. She remained undaunted, unapologetic and fiercely defiant.

In escaping, KU1, KU2 and KU3 had performed a devil's dance to perfection. She vowed to hunt them down, no matter where they had fled

* * * * *

Complete darkness with pockets of dazzling starlight had enveloped the Boston area and all of New England. It was a little after 8:00pm.

Outside, in the cold night air, even the still trees and the inhabitants of the forest groaned in apprehension.

All the world's scientific projections confirmed that the last traces of Earth's normalcy were anticipated to expire shortly before midnight tomorrow.

Chapter Forty-One

Nobility

Shortly after the failed laser strike, Mōr decided to make himself visible once more. He left the protection of the armada with the intention of confronting KU1, KU2 and KU3 in an attempt to divert their paths.

But he had to find them first.

When Shanta learned of his intentions, she was furious with him. And that sentiment would soon include the task force members and also that of the O'Riley household. All strongly opposed to his contemplated action.

The competent, but at times overbearing, Lieutenant Arak, walked briskly up to Shanta. "If our ultimate defensive weapons could not destroy KU1, KU2 and KU3, what makes Mōr think he can. Such an attempt is foolish. Nothing more than a suicide mission!".

Shanta bristled.

"I would like you to consider the fact that Mōr's motivation is born out of his love and admiration for his friends on Earth. He is driven by courage, oblivious to fear or concerns for himself. On the surface his actions and emotions appear to be reckless. I prefer to think of them as a noble undertaking!"

Mōr was fully aware that when, or if, he finally located the invaders, their singular goal would be to quickly dispatch or destroy him.

* * * * *

KU1, KU2 and KU3 had become aware of the alien intention to blast them with their massive and deadly volley of laser beams. Fortunately for them, each possessed their own intuitive guidance systems. As soon as the alien firing commenced, each had raised their protective shields and immediately became invisible. Unseen, they time-warped off in separate directions into the safe haven of deep space. They had no intention of being caught in the armada's fire control blinking red square.

The aliens, like everyone else, had seriously underestimated the intelligence and ingenuity of their opponent. And whatever Earth had in store for them, the trio firmly believed that, like the alien attack, the result would be the same. Victorious! Destiny was within their grasp!

* * * * *

Mōr believed his luck had returned when the armada's radar finally found the fast-moving invaders... back on their previous trajectory heading toward Earth. The trio believed they were now secure and immune from further attacks.

Those feelings quickly evaporated, when to their utter surprise, an incredibly long snake-like creature suddenly attacked them from behind. The moment Mōr attacked, KU1, KU2 and KU3 instantly dispersed in different directions and became invisible. Composed, they had repeated the same tactics they used to escape the aliens' laser beam assault.

Caught off guard, Mōr found himself hovering in the middle of empty space. Alone in a virtual no man's land.

But as he looked around to get his bearings, he unexpectedly became encircled by the fast-attacking KU1, KU2 and KU3. Each had suddenly popped out of the dark expanses of vacuous space, like a magician pulling a rabbit out of a black empty box.

Mōr remained in place, ready to confront the three with courage. Thinking himself invincible, he believed he only needed his colossal size and inherent strength to prevail.

The battle joined, it turned into acrobatic movements of attack and counter-attack with no one initially gaining the upper hand. But the three angry invaders, with deft combinations of lightning-fast, intricate and dazzling maneuvers, eventually outflanked Mōr.

The diving, sharp-edged trio cut and easily gutted over 4.85 kilometers or 3.01 miles of one side of his body. The wounds were deep, and though they did not bleed, the pain was unbearable…like nothing Mōr had ever before experienced.

Suddenly, he found himself completely powerless to move. Thanks to the effects of gravity, the weight of his gigantic body caused him to begin to freefall and tumble like a lifeless leaf into the dark abyss of endless space.

Mōr, with his last gasp, sent out a final distress message. As he descended into the abyss, the mental anguish of failure tormented his being. Sadly, as his freefall picked up speed, he gradually lost consciousness and the tiny light inside him slowly dimmed till there was none.

Shanta used all her guile to conceal her contempt for KU1, KU2 and KU3, as tears clouded her eyes. *I fear we have seen the last of Mōr. A hero lost forever!* The armada searching for him could not be risked, as it would leave Earth totally exposed with no defenses. Shanta decided to remain silent about Mōr's fate for now…a fate known by only three.

It was now approaching midnight in Boston.

Chapter Forty-Two

No Illusions

December 24

When Shamus and Angie met yesterday early in the morning, they decided to consider alternative courses of action. A different scheme. An "It can't be done" plan.

Shamus talked to Duffy and Michael about their ideas. Both whole heartedly endorsed their basic concepts, particularly one. It had worked a long time ago, but on a much smaller scale.

After a decision was made, Shamus left to return to Ireland, while Angie, departed for her home in the forest at St. Johnsbury.

For their plan to have any chance at success, a great amount of work had to be completed in a short period of time. World-wide fence mending, with those one disliked or found abhorrent, had to be initiated and accomplished, despite even if the very thought of doing so was perhaps ludicrous.

Reasoned purpose, had to be clearly defined and accepted by all. Similarly, total understanding and acquiescence to the underlying rational for the contemplated action was paramount.

Universal participation was an absolute pre-requisite to success. That participation would be required of the good, the bad, the believers, the non-believers, the indifferent, the evildoers and all the different fringe elements in between.

All nations and all people, such as the U.S and its neighbors in South America, Israel and the Palestinian states, Russia and the Ukraine, China and Tibet, North and South Korea, India and Pakistan, the African continent with its myriad of multi-nations…all places where tensions and strife constantly ran high… had to completely suspend and put aside *all* of their century's old animosities and differences.

Shamus and Angie always understood that they might be chasing a far-fetched miracle. Their goal to create and channel a *positive universal consciousness* might be seeking that which was truly unobtainable.

Time. It would not be their friend. By their best guess, the entire plan had to be fully operational no later than 8:00 pm that night.

Power. Not its might, but the ability of words well chosen to persuade. That was the *real* power they so desperately needed.

Ultimately the common denominator to succeed required as much luck as it did perseverance. In the truest sense, it was all for one and one for all. The continued existence of humanity hung in the balance.

* * * * *

By late afternoon, the impossible began to appear like it was possible. Most widespread resistance to Shamus and Angie's plan was gradually crumbling. In every corner of the globe, all the doubters and detractors were reluctantly agreeing to join the cause.

The exigencies of the moment were overcoming hundreds of years' worth of grudges and insidious acrimony. Reason, itself, had to be elevated to a new level.

The ultimate trick would be to get all these disparate groups to remain connected, for an indefinite period of time, to the very people they loathed or scorned.

* * * * *

For instance, one of the most difficult to persuade to join the cause were the so-called evil spirits that lurked throughout the world.

In Russia and most Slavic countries, it would be "Rusalka", a water spirit that captured and drowned its captives; in India it was the "Pretini", a vicious ghost with backward feet that preyed upon the unsuspecting; in Mexico "La Novia de

Tola" a grotesque ghost that attacked wandering children and screamed in the night like the banshees of Ireland.

For the American Indians, it was "Wendigo." A gaunt, flesh-eating creature on two legs with a horse's face that mimicked the humans it loved to eat. In Australia, it would be the "Yora-ma-yha-who", a vampire from Aboriginal folklore not unlike those described in European fiction.

In China, and most of the Orient, it was the "Monster Nian", a sea serpent with long sharp teeth and giant horns upon its head. He only left the sea to terrorize local villages on the night of China's New Year Eve.

Finally in South America there was the "Mapinquari." Horrifying and monstrous, it was a hairy cyclops with a gaping mouth in its stomach, one that it used to consume man and beast alike.

By agreement of all such evildoers, their wicked acts were immediately canceled. That agreement helped universal participation gain much needed momentum.

* * * * *

Shamus and Angie, with the help of many friends, all keeping a tenuous grip on faith, eventually prevailed on all the doubters to band together with all the believers. Fortunately, the believers were more than a substantial majority of the world's population.

By design, Shamus and Angie had included Shanta, Mōr and the O'Riley household to the inner circle in developing and implementing their plan.

Shanta had promised to use all the experience, knowledge and skills of the armada to help assure success. She and the armada would not turn a deaf ear, or reject out of hand, a whimsical idea that had once saved one of the seven tribes of Ireland's Tuatha De Danann…the leprechauns… from certain destruction.

Nevertheless, she ordered the Ryiels to be on standby for full power on a moment's notice. She so desperately wanted another crack at KU1, KU2 and KU3.

But most importantly, Shanta would be asked to communicate necessary instructions and encouragement, telepathically, to the world's entire populace and other "entities." A previously, unparalleled communication problem was solved by Shamus' inspired genius.

Thankfully, early yesterday, Mōr had been able to rapidly organize the world's aquatic population. Their spirit of cooperation and willingness to participate was unmatched. The idea of being bound together by a convincing destiny was an idea that appealed to all his sea friends.

The ocean's they lived in represented over 71% of the globe's covering, so this was an exceptionally large group of positive and cooperative participants.

And so, in its darkest hours, with Earth facing probable annihilation, all its inhabitants, without regard to color, creed, nationality, habitat, personal background or motivation, faith or lack thereof … banded together in a selfless, last-ditch effort.

An effort based on new-found hope and faith. Spirited by one of the most primal of emotions, the need to survive.

Chapter Forty-Three

The Outer Rim

In Boston it was already 8:00 p.m.

The shadows of night had slowly crept over Earth's western hemisphere. Across the eastern hemisphere the light of day grew, but was marred by a wide-spread cloud cover blocking out the Sun's rays.

Whether they lay sleepless as night descended or had risen with the dawn, the people of the world shared common companions… sheer terror filled with unreconcilable doubt as to one's continued existence. These daunting twins, merciless in their attack, wormed their way into every inch of humanity's minds and souls across the Earth.

* * * * *

KU1, KU2 and KU3 slowed their speed and paused just outside the outermost rim of Earth's magnetic field. At last, they had arrived face to face with their destiny.

Their success meant piercing the three most powerful rims of the magnetic field. Should they break through these three, a tangle of lesser strength fields, all much closer to Earth, would attempt to block their path. The invaders knew that these would not pose them any problems.

Like the scientists that studied Earth's magnetic fields, they were aware that the field's intensity or their strength went up or down for unknown reasons. What was clearly known was that the overall strength of Earth's collective magnetic field had decayed by ten percent in the last 150 years alone.

That weakening, especially included the strength of the magnetic field's three major rims. To visualize the rims, one had to think of three wide semi-circular bands extending from… and continuously connected to…Earth's geomagnetic

North and South poles. The outermost rim was the weakest in strength, the middle rim moderate and the inner one the strongest.

KU1, KU2 and KU3's plan was finalized just a few kilometers, or 1.24 miles, from the outer rim. All systems a confirmed go!

* * * * *

Likewise, over the last few days, the plan to save the Earth had been carefully discussed, considered and approved by its most erstwhile defenders: the alien task force, Mōr, Shamus, Angie and the O'Riley clan.

The armada's defensive stance had all three of its vessels positioned just *inside* the perimeter of the outer rim. The armada would form the first line of defense.

Similar to the inner workings of a clock, the exact timing and integration of all parts, regardless of size, was critical to success.

* * * * *

Shamus looked at Angie with satisfaction. "Angie's our grand scheme has reached its final stages of preparation. I have a good feeling, just as we envisioned, that our idea will provide a more than adequate second line of defense."

"Throughout the world, all land-based life will soon join hands to form an unbroken continuum from town to town, state to state, country to country. A band of sorts, albeit one that may lack any order. This meandering band of hands will weave across and through meadows, green forests and snow-topped mountains, eventually crossing a multiplicity of national borders. Invariably, our band will stretch from sea to sea."

"A similar band is about ready to be formed across the world's oceans, lakes and rivers as every member of the aquatic world seamlessly join together, either above or under the waters."

Angie laughed. "And Shamus, you'll be glad to know that I asked our new friends, the evildoers, because of their ease of movement, to bridge the small but critical area, between the Caribbean Sea and the Pacific Ocean. The nation of Nicaragua, with its very difficult terrain."

"Of course, the gentle porpoises, on each ocean side of Nicaragua, volunteered to join with these misfits to assure global connectivity. So far, all appears to be going according to plan."

"Shamus, I can't help but think that soon we will either be hailed as geniuses or ridiculed as fools."

The concept seemed simple enough. The hand-holders hoped to capture every ounce of electrical energy generated by the constantly spinning Earth's molten layer of iron and nickel. In theory, this electrical energy would be collected and stored in the body of each individual connected to the band.

That energy on Shanta's command, and with the help of positive universal consciousness, would leave each body and be launched toward one of the major magnetic rims, connecting with and enhancing the ring's own strength. The outer rim would be first.

Shanta, at 8:15 p.m. Boston time, instructed all hand-holders to make sure that their grips were secure and that their feet were firmly planted flat to the ground. This "grounding" process was designed to draw and collect as much electrical energy from the Earth's core as possible.

Shamus and Angie appreciated the fact that most in the band would be filled with apprehension. Primarily, apprehension of failure. Ultimately, this was seen as a good thing, in the sense that the apprehension might make the hand-holders more intensely focused on doing what was required of them.

It wasn't long after Shanta's announcement that each hand-holder suddenly felt a warm tingling sensation course throughout their bodies. It restored a quiet calmness, filling everyone with trust and confidence. Inspired by this strange but comfortable feeling, the hand-holders believed they were about to accomplish the impossible.

Their optimism, further enhanced by countless silent prayers to whatever Supreme Being they preferred, told Shamus and Angie that faith, not fear, would help them more than ever. Each was convinced that these prayers would all return in the form of ripples of hope and renewed strength.

Shamus believed any doubters that remained would become firm believers, even if only temporarily. Without this belief etched in the hearts of all, an omnipotent positive universal consciousness would not be obtained

It was now a little after 8:30 p.m. in Boston. Shanta began to issue her second command.

"Stand ready, hand-holders, to receive a large surge of extra electrical charges from the Earth's core, coming up through the ground and into your bodies. On my command, the electrical charges will leave your heads and immediately take on the form of a golden halo."

In an instant Shanta's words were fulfilled.

Shamus and Angie watched the wondrous sight, as each of the halos rose into the night-sky and swiftly joined together over North America, near the Pacific Ocean. There they merged, forming a single gigantic electrical bolt. The merger was nearly instantaneous, but not nearly as fast as the bolt itself, which dashed through the Earth's atmosphere and wrapped itself around the outer rim.

Shanta and the armada also observed all this with a quiet and reserved sense of gratitude!

* * * * *

"Boys, we have already disposed of that hideous monster the armada sent to attack us. Now, it's our time!" KU2 confidently shouted to KU1 and KU3. "I wonder what our mother will say when she actually meets her long, lost sons?"

In all their fury, the invaders initiated their charge toward the outer rim of the magnetic field. Their objective would be that portion that helped to protect the northern portion of the Pacific Ocean, particularly where the Hawaiian Islands were situated. The trio were each separated by only four kilometers or 2.48 miles, now closer to each other than they had ever been.

Summoning all the inherent destructive power each possessed, with their protective shields down, they attacked.

The armada's task force, including Thomas, Ellie and JP had theorized that KU1, KU2 and KU3 could not penetrate the magnetic field with their protective shields up or while invisible. Their theory was correct.

As the invaders approached the outer rim, Shanta's armada, with shield up, unleased a widely-dispersed volley of laser beams toward the trio. This volley would not be a mere "fire for effect." Shanta had ordered the armada's cannons to rotate at their base by several degrees, allowing for a wider and more generous field of fire.

Exposed and vulnerable, many of the laser beams finally had direct hits on KU1, KU2 and KU3. As the laser beams continued to pummel the three, the outer rim of the magnetic field suddenly pulsated and exploded with extremely potent amounts of electrical energy.

The fingers of highly charged electricity looked and acted like flying shards of glass from a smashed picture window, shattered into more jagged pieces than one could count. Only these shards were, at least, a thousand times more deadly.

Lynkai stood with Shanta in front of the command floor's panoramic window, arms folded quietly staring at the unfolding spectacle. "I'm completely mesmerized, as if I'm watching the turbulent dying days of a nebula. The rim's thunderous reverberations and its thousands of powerful crackling and deadly electrical bolts that erratically spring from it are marvelous."

Shanta confidently replied to her, loud enough for anyone nearby to hear. "It's obvious to any observer that KU1, KU2 and KU3 have become totally confused and disorientated, battered by the magnitude of the outer rim's power. I don't think they can overcome the blasts of the armada's laser beam attack and, I'm sure, the searing pain from the electrical charges. This combination should stop them dead in their tracts."

In a moment, Shanta's words would prove prophetic. The invaders were soon repelled. Discretion dictated that a full retreat as the only viable alternative. This was the first time the trio had ever experienced abject feelings.

* * * * *

After watching the invaders retreat, Shanta ordered the armada to bring down their shields and remain stationary behind the area of the outer rim where KU1, KU2 and KU3 had just initiated their attack.

The amount of damage to the invaders from the laser beams, if any, was unknown but assumed to be negligible. Jubilation, at this initial success, abounded in the O'Rielly household and in all the hearts of the world's hand-holders.

For now, there was nothing left to do but wait. The waiting would be agonizing, even more so than any useless attempt to try to determine what KU1, KU2 and KU3 would do next.

Shanta had a brief opportunity to revisit her silence in not revealing to the unsuspecting that Mōr had, undoubtedly, been lost in the obscure depths of space.

I will do so soon, but now is not the right time!

* * * * *

The three invader's first attempt to penetrate the outer rim was easily rebuffed by the sheer strength of the armada's shields, the wide-spread magnitude of the laser beams and the unusually powerful electrical charge of the outer rim itself.

Each of the invaders looked down with disdain at Earth's hand-holders. All of them clearly elated as a result of their successful attempt at reinforcing the outer rim.

KU1 screamed as loud as he could in anger. "I vow that the confidence of these ruffian hand-holders will be misguided. Their initial success short lived. Guaranteed!!!"

For KU1, KU2 and KU3 a new game plan had to be devised. One of the first things they decided to do was to conjoin as one. By conjoining they increased their mass more than three-fold. This would enable them to increase their ultimate speed of attack to an improbable high far in excess of the speed of light. Such a merger would humble Einstein's famous equation.

Jack, in total dismay, watched as the invaders conjoined on his computer screen. "Shamus, this conjoining is the worst of the possible worst-case scenarios possible. We need not consider the 'end of days'…rather the likelihood of the permanent end of time in Earth's corner of the universe."

Shamus placed a hand on Jack's shoulder. "Keep the faith, my friend. Keep the faith!"

The invaders once again struck the armada's shields and the outer rim with deafening and ferocious force. But, despite their increased mass and speed, their second strike at the outer rim was again repulsed. Repulsed for the same reasons as their first attempt.

Foiled again, the invaders retreated filled with rage. This time deep, far deeper than before into the empty darkness of space.

Serious introspection, sprinkled with a high level of anxiety, galvanized their every thought. But it was not too long before KU3 brokered a new idea. Why not add substantially more distance to the point of attack? Its calculations indicated that by doing so, their speed and momentum might possibly increase *ten-fold.*

KU1, KU2 and KU3 patiently hovered for a while as they prepared to attack the exact same spot on the outer rim. This time they would not only be conjoined, but emboldened with ten times more momentum and velocity.

The armada's shields were easily breached and, at the same time, their ability to maneuver with untold speed enabled them to avoid almost all of the armada's laser beams. Their penetration of the outer rim, though slight, caused an area of the rim to take on the shape of an indented half-egg.

Momentarily stuck in this indented half-egg, the invaders were immediately covered and incessantly jolted by super-charged electrical shocks. The electrical shocks proved to be extremely strong and lasting. Any attempt to endure their effect, for any length of time, meant imminent disaster!

But with great effort, they were able to wiggle free and extricate themselves from the half-egg. Necessity was unyielding. It required that they retreat once more. Disappointing, yes. But they were not disillusioned!

The trio moved with newfound confidence to an even further distance than before, into outer space. After taking a few minutes to recharge, they proceeded to recalculate the principal factors involved in the equations that affected their next attack. *Distance. Mass. Velocity.*

In increasing their distance from the target with conjoined mass, the three would attack, in theory, at the highest velocity that they could ever hope to physically achieve. This, they thought, would make them an unstoppable force. Practice would indeed, make perfect!

How bizarre it was that after all they'd been through, their future now rested on the penetration of an indented half-egg. Ludicrous, but nonetheless a serious obstacle. With mild concern the trio thought to themselves: *This half-egg better be fragile!*

This time, all their calculations proved correct. KU1, KU2 and KU3 exploded through the armada's assorted defenses and the indented half-egg portion of the outer rim like a humongous star burst. The electrical tails of the rim, spent, followed them for a considerable distance.

The hour was now 9:45 p.m. in Boston.

* * * * *

The armada began to move with lightning speed to an area behind the second rim, aligned with the exact point of the outer rim's breach.

On the command vessel, Shanta remained dumbfounded. She knew that what she just witnessed might become a most unwelcome routine. A mixture of sadness and anger flowed through all the fibers of her being, as she confronted and reviewed the futility of the armada's defenses. *I know what failure is, but this one has a deeper meaning to it. I feel as if I've slipped off a craggy cliff and am plunging into my own personal hell.*

Her only solace was in the fact that the outer rim was the weakest of the three. They still had two chances left to salvage the situation. More than one, yes… but only two!!

On Earth, a shudder of disbelief ran through the O'Rielly household and all the hand-holders when they learned of the breach. In a little more than an hour, the outer rim had been penetrated and the globe's positive universal consciousness scattered in disarray.

While trust and faith had begun to rapidly wane, Shanta's mind was now multi-tasking…focused on other startling events that had been set in motion.

Chapter Forty-Four

A Resurrection

Only Shamus, Angie and Shanta had received Mōr's last distress message which he had sent just before losing consciousness. Each had thought it best to keep their knowledge of the fateful message to themselves so as not to raise false hopes on Earth. The three immediately put into place the only viable search plan thought feasible.

The help of Lynkai and the armada's task force was deemed to be absolutely indispensable for the plan to succeed.

Shamus was placed in charge of one of the armada's most advanced and fastest vessels, loaded with medical specialists and other necessary supplies. After picking up Shamus on Earth, the vessel time-warped to an area in deep space where Mor was projected to be located by the armada's task force computers.

"Angie, Shanta and Lynkai, we've arrived at the coordinates given, but we do not see any signs of Mōr," Shamus reported.

Lynkai was the one to respond. "One moment, Shamus, we'll immediately perform additional calculations and projections."

Once provided with the new coordinates to an uncharted area deeper in space, Shamus set the vessel's controls at max time-warp speed. Though it seemed like it took an eternity to arrive at the new destination, it was actually only a matter of twenty-four seconds.

One slight turn…there! Unexpectedly, Mōr's long body appeared like a bizarre apparition. Mōr could be seen drifting in a listless state in an unexplored region of starless, dark space.

Instantly, the medical teams suited up for the environs of space and departed the vessel, racing to repair Mōr's multiple injuries. Shamus requested Captain Myka to assume control of the vessel and left the command center to begin suiting up for his spacewalk. He was given the difficult task of injecting Mōr with special drugs, prepared by Ireland's leprechauns, to overcome his loss of consciousness and predictable lack of strength.

Shortly after the first of three injections by Shamus, the tiny light inside Mōr began to dimly glow. Hope! *These drugs could actually work.* Ten minutes later, the second shot caused tremor-like movements in portions of Mōr's long body, followed by low grade rumblings. Believe! Shamus, Angie and Shanta always kept the faith.

At the appointed time, Shamus administered the experimental…first-of-its-kind…third injection. Tremendous patience was now called for as everyone had to wait for its effects. After many anxious minutes, hope and faith would not be denied. Mōr abruptly sprang to life. Soon to return to his old rambunctious self.

No one would ever know the actual amount of time it took to revive Mōr, but Angie and Shanta knew this resurrection had to be an all-time record in the making.

By the time Mor returned to the armada, under his own power, he was as good as new and felt stronger than he ever had. *My friends the leprechauns, indeed, have unsurpassed magical powers. I can't wait to thank them in person.*

Little did he know!

Chapter Forty-Five

The Middle Rim

KU1, KU2 and KU3 having completed the breach of the outer rim, returned to deep space to recoup their stamina.

Next up would be the attack on the middle rim.

Shanta's armada planned to defend the middle rim with the same strategy used to defend the outer rim. Hoping that a repeated pounding of the invaders would eventually take its toll.

She would implore all the hand-holders to double their efforts to store as much electrical energy as possible. When the time came, they needed to concentrate it with all their might into another very powerful electrical bolt.

* * * * *

It was not enough. KU1, KU2 and KU3 easily breached the middle rim after only one attempt. The time in Boston was 10:30 p.m.

Shanta was a professional, both a stateswoman and a tested commander. She had seen conflicts, made decisions that no one in the universe could have. And yet, she had to reluctantly admit the obvious. The efforts of the armada and the people on Earth, together or otherwise, had failed quiet miserably in all respects.

The invaders were off the charts. Their complete departure from anything closely resembling normality was baffling, confusing and frustrating, not only to her, but to every one of her people. It was humiliating, witnessing such immense losses to deviant tactics.

And for those assembled in the O'Riley household, the realization that it only took the invaders 45 minutes to pierce the middle rim was terrifying. Tumultuous fear swarmed throughout the home, like hornets feasting in a warm, budding flower bed.

* * * * *

KU1, KU2 and KU3 basked in glory as they now lingered back in the deep recesses of space. As they rested, each strangely experienced the human feeling of adrenaline rushing through them.

Human or not, their thoughts turned diabolical.

Only one last hurdle to overcome! All their projections yielded favorable results with a straightforward attack at maximum speeds.

The outcome… a joyous meeting with their mother.

Chapter Forty-Six

The Inner Rim

There is a plurality of the world's population that believes we do not control our own destiny. Shamus and Angie were of a contrarian opinion. To them, even the slightest hint of predetermination was foolish gibberish. One had to own their life's course, with their own hands, their own actions or lack thereof, controlling their own present and future.

At 10:45 p.m., standing on the O'Riley back porch, Shamus touched Angie's shoulder.

"We've got a problem!" Shanta just informed me that she was checking the command's vessels monitors, when she noticed an enhanced depiction of the current position and status of all the Earth's hand-holders."

He pulled out the laptop he'd so kindly been given.

"Here, I'll show you. Look closely at the borders between the unfriendly countries just before KU1, KU2 and KU3 breached the middle rim. You can see what she is concerned about."

Angie noticed right away that in almost every country the electrical strength of the hand-holders was bright and strong. This was true except for the many areas around the globe where the *borders* separated antagonistic countries.

At these hostile borders, the hand-holder's electrical charges were non-existent, extremely weak or not properly connected. Obviously, due to a lack of trust in one another.

A sampling told the entire story: the borders were extremely weak where China met Tibet, Russia met the Ukraine, India met Pakistan and Israel met the Palestinian states, the list went on and on. Drastic action was needed!

To that end, leprechauns and all of Angie's family members from all over the world, promptly descended on the troubled border areas to encourage committed participation. Each would be fluent in any language spoken.

Shanta, a bristle in her voice, let it be known to all the hand-holders on Earth that the situation was extremely dire and getting much worse by the minute.

"As you all know, our last hope to save Earth is to defend the inner rim. All our efforts to prevent the outer and middle rims from being breached have failed. The reasons for our failure do not matter. What does matter is what we all do *together* now."

"Without positive universal participation and unconditional trust by all, for all, and in all… success will not be possible. Tomorrow's dawn is now elusive, perhaps even an afterthought, just like your existence. Humanity and all life on your planet are close to a doomsday scenario."

"But together we will not fail. Together, with trust in each other, we shall… indeed must… produce a more favorable outcome!"

The resolve of the leprechauns and the white eagles would not be denied. In a few precious minutes, they circled the entire globe dispensing their magic. The command vessel's screen now showed well-illuminated and fully connected hand-holders circling every inch of the entire globe. For the moment, at least, trust in each other was restored!

* * * * *

While the attempts to solve the border problems were on-going, Thomas, Duffy, Michael, Jack, Ellie, JP, MJ, Sean and Lynkai had been in serious discussions relative to another matter of urgent concern. The present strength of the electrical charges generated by the hand-holders, in the final analysis, would not be sufficient or strong enough to fend off the invaders.

An idea struck JP. He stood and explained: "To generate stronger electrical charges, the speed of the Earth's spinning has to increase substantially. A self-driving dynamo process is needed. It needs to be of unimaginable strength and robust proportions. The issue is how can we produce such an effect?"

The lack of time to discover such an out-of-this-world solution was crushing.

Ellie, for reasons unknown, was more than surprised when she conjured up an image of her grandfather. The apparition stood in front of her and whispered a message. In a flash, her grandfather disappeared and she broadly smiled to herself. Nothing happened in the universe without a reason.

In a heartbeat, Ellie asked Lynkai about her thoughts as to whether or not Mōr could produce the affect needed. She took a breath. "It would depend on a number of factors, but…"

* * * * *

KU1, KU2 and KU3's moment of truth was at hand, as they finalized preparations for their advance on the inner rim. They knew the armada's defense was likely to unleash the most powerful laser volley yet. For that reason, when they went through the openings previously created in the outer and middle rim, they planned to be invisible, with shields up and travel separately, a safe distance apart.

This would allow them to evade the onslaught of the laser beams. Though the inner rim was the strongest in terms of electrical charges, they still had no real or imagined thoughts of impending danger. Their determination and strength were thought to be more than enough.

All was ready.

* * * * *

It was 11:05 p.m. in Boston.

Somehow, to the utter dismay of those monitoring the situation, all hell had once again broken out at the world's unfriendly borders. It immediately became clear that the collecting and sending of electrical charges through these hand-holders would be woefully inadequate.

It was time for a certain group to right their every wrong. To do so they needed to do the very things that caused the wrongs in the first place. All the world's evildoers were summoned and requested to take drastic action to scare the wits out of the problem border's hand-holders, in the hopes they would correct their attitude.

One such occurrence was emblematic of them all. On the India-Pakistan border, few of the hand-holders were trusting each other. The two countries had long shared an antagonistic and malevolent history. Now these hand-holders were verbally abusive to each other, waiving raised arms and fist threatening physical harm.

Suddenly the monster, that both cultures knew and feared the most, appeared in multiple apparitions to the entire border. The dreaded "Pretini", a particularly vicious and cruel ghost, warned them all. "If you do not stop your foolish antics, I promise you…and all that you love…will perish at my hands before the hour reaches midnight."

Normally, to expect a positive result from this conflicted group would be uncertain at best and foolhardy at worst. But in the end, horrifying and frightful renderings of the coming apocalypse, by one's deamons, turned out to be extremely helpful.

Taking responsibility for one's errors and wrongful actions was an honorable thing to do…even for evildoers.

*****

It was nearly 11:15 pm in Boston.

KU1, KU2 and KU3 had already hopscotched their way through the openings in the outer and middle rim. Once they arrived at the inner rim, they couldn't help but notice that the rim appeared to be much more luminescent than the other two rims. A minor trifle.

Destiny awaited them. And destiny, cruel as it could be, demanded that they punch their way through the third rim.

*****

Precisely five minutes earlier, all the aliens had crossed their arms across their chest and closed their eyes for what seemed like an eternity. In fact, it was only thirty seconds or less. Their surreal attempt at obtaining a positive universal consciousness was more effective and much stronger than Earthlings could ever achieve.

Shanta watched from the command vessel, as Mōr disappeared in an enormous swirl. He was on his way to wrap himself completely around Earth. Once that was done, he would become invisible. Thanks to his alien friends, he was now almost ten times his previous size, imbued with the power of fifty Ryiels.

With a look of certainty, Mōr knew he could make the enormous mass of the Earth spin faster and ever faster. What he didn't know was whether he could keep the delicate balance of its axis and orbit stable.

Back in his filled-to-capacity study with everyone glued to the computer screens, Jack nudged Ellie and asked her to explain Mōr's predicament.

"One has to think of the problem in common-sense scientific terms. If a spin-top, with *no* outside guidance or control, has a fixed orbit around an object like our Sun, and you became able to suddenly spin it faster… then even faster… its axis and orbit would *undoubtedly* shift or wobble. It would fall out of orbit."

"As you can imagine, such an event, a wobbling axis and out of control orbit, would be both disastrous and fatal for Earth and its inhabitants. A disaster that would make our problem with KU1, KU2 and KU3 look like a sore thumb."

"But there is some good news. Thomas, has informed me that he, Lynkai and the think-tank have found a way to resolve this issue."

* * * * *

Shanta's voice told Earth, at 11:18 p.m., that KU1, KU2 and KU3 were on the doorstep of the last and most powerful rim of the magnetic field. Failure was not an option!

The world's hand-holders, their nerves entirely tattered and frayed, lifted their eyes toward the heavens in one final and desperate attempt for help.

The Earth, indifferent to the cosmic threat overhead, somehow remained fixed to its normal orbital path. Its rotation, however, had temporarily stopped. This produced two effects: one intentional, one accidental.

As intended, the instant that the maximum possible spin rate was achieved through Mōr's efforts, Earth's core produce a monumental burst of electrical energy larger than any witnessed in the billions of years since its formation. This electrical energy arched into and out through all the hand-holders.

So strong was this burst of core energy that every halo generated by a hand-holder, increased in size tenfold. When these abnormally large halos were seen, it unified the people of Earth in a way they could scarcely imagine.

The unintentional effect was, that the surface of the Earth, coated in the light of an ever-present Sun, shown so bright that all night faded. Where the laws of the universe dictated that one half of the Earth lay dormant as the other rose, where one slept as the other woke, the events of the present demanded that all people of Earth be privy to a night without darkness, one full of light.

And on this day with all the hand-holders filled with faith, hope, trust and even love… their halos were channeled into not one, not two, but seven gargantuan electrical bolts which then flashed at the speed of light to the inner rim. So bright were the bolts that even Shanta had to shield his eyes.

* * * * *

The invaders neared the boundaries of the inner rim, grateful that at this point they had been successful in warding off the armada's defenses. Strangely though, each felt a sudden increase in their individual weight or mass despite the fact they had not yet conjoined.

Thought curious for the reason, they quickly attributed it to good fortune. There was no time for introspection.

More mass, produced more velocity, and more velocity produced more destructive powers. It was an axiomatic scientific principle, that the three took great pleasure in recognizing. And as they did - "Gonzo"- they penetrated half-way through the thick and broad expanse of the inner rim.

* * * * *

Shanta and the armada's entire crew watched KU1, KU2 and KU3's approach. Shanta's lips whispered a silent prayer to a Being she did not know or had ever contemplated. What she did know, she kept to herself.

* * * * *

At precisely 11:24 pm Boston time, an ominous thick kelly-green cloud of unknown origin formed and covered the exact area where KU1, KU2 and KU3 were attempting to pierce the inner rim. Surprisingly, not even the armada's radar instruments were able to see through this mushrooming green cloud, whose circumference began to emit a bright light.

As the seconds ticked away, the armada was still unable to see through the now expanding and obfuscating green cloud. Its light increased in intensity to where it became blinding, another formidable barrier to visual observation.

Seconds turned into minutes.

The cloud emitted no noise or signs of anything atypical transpiring.

Minutes soon trudged into clips of five minutes or more. Nothing!

Shanta was usually one who was unmoved by joy or grief, the mundane or the sublime. She always preferred the reality of the moment. This time though, acceptance of her present reality wanted to escape more than her overly objective and rational mind.

The minutes now grudgingly gave way to segments of more than ten minutes. The waiting slowly becoming almost unbearable. On the armada and on Earth, tenson was pervasive. Mental stress had adopted a child named "frazzled."

Time stood still.

To the surprise of everyone on Earth, the green cloud's wide-spread blinding light blocked the stars and the heavens from view. It was a maddening time.

* * * * *

Inside the cloud, KU1, KU2 and KU3 were just as frustrated and perplexed as Shanta and the armada. Nothing was happening as planned. As the three had started to pierce the inner rim, multiple electrical bolts of immeasurable strength, overwhelmed them and stopped them dead in their tracks.

These electrical charges then sought out every fissure or crack that existed in the invader's structure. Once found, they wormed their way deep into each crevice and seemly remained stationary. For the three, this was difficult to comprehend much less understand.

Too late, they would each learn that their added weight was caused by the accumulated water in their fissures or cracks. This water had been melting, heating and then expanding. This led to the weakening of the very integrity of their structures.

Another onslaught of electrical bolts, twenty times more powerful than the first they encountered, began to effortlessly flow through every micro-inch of their bodies. Though slowly at first, soon each fissure or crack in their respective bodies grew wider and wider, faster and faster.

Just when KU1, KU2 and KU3 thought it couldn't get much worse… it did! The growing expansions of their fissures and cracks were infiltrated by an unexpected and new plethora of stronger and much more deadly electrical charges.

Now the three, painfully remembered a universal truth. Ordinary water becomes a *conductor of electricity* when charged ions and innumerable impurities are dissolved in it. Such was the water in the invader's fissures or cracks.

KU1, KU2 and KU3 had caused a great amount of mental anguish in this part of the Solar System. Now it was their turn to endure the same.

As the abundance of increasingly powerful electrical charges continued to surge into every fissure or crack, the strain to keep their structures together became unbearable. Each of the interior structures of KU1, KU2 and KU3 began to fracture a million times over, every inch of their mass consumed by massive electrical shocks.

It would not be long before one of the immutable laws of nature prevailed.

KU1, KU2 and KU3 now appeared as objects aglow, covered in tenacles of flashing and jumping electricity. Each had reached critical mass… the point of no return.

At 11:36 pm, each spontaneously exploded.

The resulting explosions and shockwave, filled with thousands of mile-long electrical bolts, immediately dispersed the mysterious thick green cloud. Spewing from each explosion were tiny and harmless remnants of the three invaders that would be swallowed up in the vast emptiness of the Universe's dark space.

Forever and ever. Dust to dust!

While it seemed like an eternity, only twelve minutes had transpired.

KU1, KU2 and KU3's dreams of a future with their mother dissolved in a catastrophe of their own making. Certainly, one they could never envision.

* * * * *

Shanta was grateful that Thomas and Lynkai's respective groups were able to think of a method to increase the strength of the Earth's electrical charges. Later, it would be called a miracle that Mōr was able to spin the mass of Earth at such unheard speeds without disrupting its axis or orbit.

In truth, it was more likely due to a progression of thoughtful insights from the O'Riley group as well as the armada's think-tank. The solution had a circuitous and difficult birth.

It was a given that Mōr had to wrap himself completely around every inch of the Earth's surface to create the spin rate necessary to produce the ultimate speeds obtained. In doing so, he formed a protective cocoon around the Earth.

It was left to Lynkai and Ellie to then realize that if the cocoon could move Earth and its axis in a stable fashion, it might also be able to do the same with

Earth along its normal orbit. One problem. Mōr couldn't see as his head, to reduce friction, had to be inside the cocoon.

Enter Shamus and Angie, who came up with the idea to take the brightest stars in the heavens and temporarily place them *on* the path of Earth's normal orbit. Next, they designed an internal guidance system for Mōr to guide him over these special stars.

It all worked perfectly. The necessary spin rate was achieved without the Earth's axis moving and stars lighted the planet's stable and safe journey on its traditional orbital path.

No wonder the world would always think of this as "Mor's grand miracle."

For the O'Riley clan, just like the rest of Earth, the explosion that many witnessed in the heavens did not immediately signal success. It would be Shanta, minutes later, that delivered the euphoric news that brought world-wide relief and untold bliss.

And as the magnitude of the moment sank in, tears of joy fell everywhere. The world's population sank to their knees, spent from physical and mental exhaustion. All eternally grateful that "Someone" had helped them.

Shanta and Lynkai, together shared the relief of a relaxing moment. Lynkai, with a perplexed expression said, "Shanta, I have no idea from where that green cloud materialized or what it is?"

"Nor do I, Lynkai." Laughing, Shanta coyly continued. "The only answer that occurs to me is that Shamus' leprechauns must be responsible."

"Really? I don't believe their powers extend that deep into the cosmos."

"Perhaps we may never know the answers to your question. And, I'm OK with that. I'm more than happy with the final result. How do you want to celebrate?"

In Boston, it would soon be 12:00 a.m. Christmas Eve. Fond memories of past Christmas Eve's would visit many of the world's population. Regrettably for some, those dream-filled memories fell apart by dawn.

But for the children everywhere, wondering was now their only occupation. Despite all that happened, they wondered whether Santa Claus was still coming?

* * * * *

The night air was bitter cold, whipped by sharp winds gyrating in every direction. A brilliant fill moonbeam cast a broad swath of dazzling, golden light over the snow-encrusted-land of New England.

At last, peace and relative calm had finally found their way and settled in on this Christmas Eve night.

* * * * *

Out of nowhere, a few were eerily stirred back to reality. Their tranquility interrupted, shocked by a tearful and desperate cry for *help*. Why did trouble always consider calm and peacefulness to be an illusion?

Chapter Forty-Seven

The Sleigh

11:45 p.m. The cry for help came from Kristy at the North Pole. It was heard and answered only by Shamus, Angie and Mōr.

Moments ago, Rudolph had suddenly collapsed. He now lay on the ground, unable to move, coughing badly and exhibiting shallow breathing. Kristy had laid a warm blanket over him as Finney and the elves placed cold compresses on his forehead and body in an attempt to cool down his high temperature.

Kristy watched very carefully but, down deep in her heart, her maternal instincts told her Rudolph would never fly tonight. Finney was of the same opinion. Rudolph was now sweating profusely growing weaker by the moment.

Thankfully, the other eight reindeers...Dasher, Dancer, Prancer, Vixen, Comet, Cupid, Donner and Blixen ... were feeling fine and already hitched up to the sleigh.

Ready to go.

But none of them had Rudolph's intuitive skills of guiding Santa's sleigh and honing in on the addresses of where all the good children lived. Her computer-like system was loaded with eons of irreplaceable data.

There was precious little time left to find a replacement for Rudolph, as Santa carefully rechecked the distribution of weight on his fully loaded sleigh.

In less than a minute, Shanta and the armada's crew were brought into the loop to help solve the problem. Before one could spell hallelujah, Mōr, in a radiant ball of bright green light reduced himself to a fraction of his true size and turned into a replica of Rudolph. Shamus' magic providing a huge assist.

A red nose, antlers, puffy white tail, four legs and all. The only essential Mōr lacked was Rudolph's internal guidance and honing system.

Mōr was downright giddy with his new appearance. *I can't wait to be a part of Santa's awesome adventure. How blessed am I?*

If he only knew the half of it!

Shamus grabbed the reins that now hung from the body of Rudolph's replica and led him to the sleigh. There he was hitched up to the other reindeers, at Rudolph's customary position of front and center. As Mōr walked, he sensed the uneasy stares and general nervousness of the other reindeers.

Once in position, Mōr, with his red nose brightly aglow did a playful step-dance prance to give the other reindeers and Santa a laugh or two. It was also an attempt to instill confidence in them that this "rookie Rudolph" was up to the task.

While Mōr did his dance routine, Angie whispered to him and Santa that the armada had been able to successfully download and transfer all of Rudolph's original guidance and honing abilities. They were now installed in Mōr's internal computer.

Santa was elated when he heard the good news. No longer would there be any anxiety over directions and honing, the hardest part of the sleigh ride.

He kissed Kristy goodbye, jumped on the sleigh, grabbed the reins and hollered out a few commands.

Instantly, the rookie Rudolph and his team sprang into the star-filled night, and began to ride on top of the light of the full moon's wide and sparkling moonbeam. This Christmas Eve moonbeam did not, and never would, obey the laws of nature.

In one continuous motion, the moonbeam would completely encircle the Earth many times over in mere seconds. For all those riding on the moonbeam, time, as others understood it, did not exist.

Santa's sleigh ride around the world would be typically magical and thankfully uneventful. And, wherever good children lived, thoughtful presents and many wished for toys were now piled high.

On their return to the North Pole, Mōr was greeted like a hero. With tears in his eyes, Mōr did all he could to deflect the praise. He reminded them all, that if not for Shanta, the armada and many other great friends, their successful Christmas Eve sleigh ride would not have been possible.

December 25

On Christmas morning, the entire armada for the first time in their lives would gather around gayly decorated Christmas trees. Each tree's wide branches covering a mountain of special presents delivered from the North Pole.

Later, it would be difficult to describe the tremendous amount of joy and happiness this brought to Shanta and her crew-members.

* * * * *

Back at the O'Riley household, Santa had clearly outdone himself with unique and wonderful presents for all. Jack had made his famous eggnog and was handing out Irish crystal glasses filled with the tasteful drink.

To everyone's surprise, there were tiny boxes, wrapped with sparkling green paper and tied with delicate red bows, hidden far under the tree. There was a box for everyone, with a note that said it was to be opened last and by everyone at the same time.

Thomas sat on the floor next to JP, Ellie, Sean and MJ near the fireplace. When Thomas was handed his tiny box, he started to sing parts of the words from Clement Moore's famous poem:

> *When, what to my wondering eyes should appear, but a miniature sleigh and nine tiny reindeer. With a little old driver so lively and quick, I knew in a moment it must be…*

As he opened his box, he abruptly stopped singing. Ellie, JP, the children and the rest of the household were awestruck when they opened their own boxes. Wide smiles filled the room as everyone exchanged knowing glances.

Warm feelings of appreciation were contagious, as Shanta's Christmas card… attached to each box…was read. "My gift to you, with all my love and admiration, to memorialize our triumph." The box contained the dust particles left from KU1, KU2 and KU3's spectacular ending.

Thomas, over afternoon coffee, would describe this moment as a uniquely "spiritual" event. His metamorphosis was nearly complete.

*****

In an alcove off of the family room, stood a tall intricately carved grandfather clock with long, dangling gold chimes. It was a precious heirloom handed down from generation to generation in Jack's family.

The chimes were scheduled to ring every hour on the hour. The hands on the clock had just clicked, momentarily stuck on 4:00 p.m.

Miraculously, for the first time since December 15th, the chimes rang out. Only Jack heard the soft tinkle of the chimes, fully aware of what this foretold. He smiled and looked deeply into the fire's embers.

He did not search for answers…just for gratefulness.

Chapter Forty-Eight

Heaven Can't Wait

The inscrutable Mōr would continue his adventures throughout the universe, confident that his knowledge of the secret of secrets would enable him to be a positive influence wherever he may be needed.

His experiences in traveling the universe and his observations of its sheer wonderment led him to the belief that all that was, all that is, all that will be, was the result of an unseen, intelligent consciousness.

A living, boundless, ethereal "Presence." Its ubiquitous and gentle imprint ever-present in the constantly expanding universe. Mōr believed the crystalline word…Presence…was a far more accurate description of its reality than the term Deity or Supreme Being. Mōr encountered many advanced civilizations that had ultimately arrived at the same conclusion.

Except on Earth: with its history of prolific religions and attendant multiplicity of competing Gods. Each claiming to be the sole beacon of the *one* true faith… the sole purveyor of the *ultimate* truth…the sole possessor of the *one* true God.

Exclusion, rather than *inclusion*, had become religion's bulwark. Exclusivity giving rise to the insidious specter of intolerance, where contrarian opinions and beliefs were routinely stifled and eliminated

Since the age of reasoned inquiry, religion had battled science for dominance, vigorously attacking any scientific inquiry that cast doubt on the validity of their spacious doctrines.

Religions, to survive, require that all their resolute dogmas and doctrines be couched in the nebulous concept of *faith.* Requiring one to trust in that which is amorphous, unseen and not tangible.

Science is the absolute opposite of faith. It categorically rejects religious conjectures, superstitions, and its whimsical doctrines. Science demands exacting physical proof for every discovery. And once an unknown is proven to be fact, it acquires a singular truth. A truth that may be subsequently changed as science's thirst for clarity advances.

When Mōr viewed the Earth's religions, it did so bewildered. The ruse of most religions continued unabated and the reason for same was not profound. Humanity, in general, *is* misguided to think it is enlightened!

What of religions whose betrayals included amoral wars, cruelly fought for over two thousand years. All justified in the name of *their* one true God. What of centuries-old insolent hatreds that continue to be resolved by acts of unimaginable atrocities. Enlightenment?

Where in the definition of enlightenment do we find the storied indifference to the plight of the poor that continues or the flourishing ethnic and religious intolerance that rages in every corner of the globe?

A life-giving planet has been, and is, treated with great disdain with, among other things, the endless pollution of its air and waters. Man's self-destructive assault on Earth's environment threatens the very viability and survival of future generations. Acts, that a minimal amount of rationality would deem selfish and sacrilegious.

The examples of humanity's continued gross ineptitude were endless. None synonymous with enlightenment! But as grim as things might seem, Mōr began to smile.

To the surprise of even awe, Mōr's old friend, Sir Thomas, had just proclaimed to Duffy and Michael on a recent visit to Ireland that: *Your God is my God…and my God is your God…all inextricably bound as one Presence. In this regard, I have at last resolved my long dialogue with my dear friend Mōr.*

What an interesting assertion for a former agnostic and world-renowned physicist. What caused this transformation? One thought suddenly cascaded in Mōr's mind.

The one area where science struggles is explaining the happenings of the unexplainable… miracles that manifest themselves in every day physical reality?

Mōr understood that miracles were everywhere, in a part of all that exists and in each of us individually! They are in the natural order and the inexplicable of the Universe itself:

> It's when you see [your] world in a grain of sand.
> It's when you see a heaven in [the colorful petals of] a wild flower.
> It's when you sense a Presence in a sunrise or a sunset… and in the birth of a newborn.
> It's when you hold infinity in the palm of your hand.
> And it's when you *believe* in miracles…that *faith* reveals the unseen Presence.

Epilogue

She loved to playfully take the form of the species that lived in the dimension presently of concern.

Her flowing robes shimmered as if they were woven from stardust. She walked gracefully, with a distinct elegance, on the silky beach sands of time. Appearing much like her favorite actress, Audrey Hepburn.

A gentle breeze twirled the long strands of her auburn hair. She now gazed out on a vast ocean without a horizon, whose waters were filled with majestic waves.

The final vestiges of a wide golden moonbeam were receding from the surface of the roiling ocean, slowly fading like the last wisp of a vanishing ghost. A brilliant pink-yellow sun had just begun its towering rise in the east.

One of Her companions, the delicate and pretty angel named Snowflakes, struggled to keep pace with her deep footprints. Snowflakes had always been instrumental in any situation that had a Christmas theme.

Snowflake's tag-a-long friend was none other than one of the "love angels" who inspired JP and Ellie's hearts to dance as one. Ryan couldn't wait to ask a few questions about certain items that remained unresolved for him.

Whenever Ryan talked to the Presence, he thought in terms of 'She'. While Snowflakes was just the opposite…thinking in terms of 'He'. Nevertheless, each understood the uniqueness of the miracle …there was but one Presence.

Ryan, questions ready, boldly fired his first one.

"I thought KU1, KU2 and KU3 were all powerful and indestructible."

Her voice, filled with a soft touch of merriment, responded. "Ryan, you should understand by now that the universe, since its existence, has always had certain immutable laws.

"KU1, KU2 and KU3, like all rock-type objects, had minuscule fissures or cracks running throughout their bodies. When the three struck Triton, Neptune's largest moon, those fissures or cracks opened and inadvertently collected a fair amount of moisture from the moon's icy surface."

“As the trio jetted through the Solar System, the heat and cold they encountered caused the collected water to continually expand and contract. But the straw that eventually broke their backs occurred in between their multiple attacks on Earth’s magnetic field. It happened when they continually entered Earth’s heated atmosphere and soon thereafter darted back into outer space’s sub-zero cold.”

“These *extreme* alternating conditions created the repeated expansion and contraction of KU1, KU2 and KU3’s fissures or cracks. In essence, their impetuous decisions made them vulnerable and are just as much responsible for their demise as are other factors.”

“By the way, the inspiration of increasing the Earth’s spin-rate to create exceptionally strong electrical bolts was sheer genius. I now believe that when I passed out intelligence for those in the O’Riley household and the armada’s think-tank, I must have given them a little more than I first thought.”

“You should also know that I had nothing to do with that particular inspiration…though my dear friend Papa surely did.”

* * * * *

“Can I ask you one or two more questions about…”

“Of course, Ryan. I left to Snowflakes the manner in which Shamus made his initial contact with Mōr. Her choice of the three Christmas stars to guide Shamus to Mōr was clever…as well as her use of them as the *voice* that delivered the riddle to Sean, MJ, JP and Ellie. I liked the idea so much that, in the future, I intend to use those three stars as a part of my personal messengers.”

“And if you had to ask that question, then you probably didn’t figure out the riddle.”

“Sure, I did. That was easy. It was MJ and Thomas!”

“Ryan, your analytical abilities need some work. The ‘young one who warned’ was the youthful astronomer Kim Alosco with her magazine article. The ‘wisest among them’ was obviously Mōr, with all his innate and artificial intelligence.

* * * * *

The three walked quietly for a period of time, until Ryan decided to jump into the ocean and body-surf some of the huge waves. The bigger, the better, for this ruggedly handsome angel. More than a little prone to folly.

He and Snowflakes decided it best to sit on a dune and assure that Ryan didn't get dismembered, or worse drink half the ocean.

"Snowflakes, I must admit that you and Ryan were exceptional when it came to JP and Ellie's adorable holiday love story. I know how happy they are with the decisions they've made. I can see them both right now.

"Here, you can look too, as He stretched out his hands and magically made a viewing screen out of thin air."

* * * * *

JP and Ellie were walking, hand in hand, on the long stretch of beach in front of their home, admiring the three day-time moons that hung just above the ocean's horizon. Each moon was circled in wide stripes, seemingly painted with the brightest colors in a rainbow.

The ocean's medium waves were gently awash on their bare feet. It was hard to imagine that they had already lived here for over two years.

In that time JP had finished his doctoral program, a New Year's Eve wedding had taken place at the O'Riley homestead and young Thomas Jr. had been born. Shanta and some of her friends were now enthusiastically babysitting for the precocious one year old.

JP and Ellie decided not to return to teaching. Their main focus became leading an elite research group that was unraveling the intricate and unsettled mysteries of dark matter, dark energy, black holes and nebulae, with a specific emphasis on the *origins* of the Universe.

JP as he walked with Ellie, kicked the sand with his toes and gently squeezed her hand. "I want you to know that my heart is bursting with joy and love for you and Thomas Jr. I can't imagine a moment in my life without you."

Ellie warmly hugged JP's arm and looked up at him with sparkling eyes full of love. Hon, you are not alone with those feelings. I get goosebumps all the time when I think of the three of us."

A rogue wave steeply curled and suddenly crashed, sprinkling JP and Ellie with cool ocean spray.

"JP. Remember to pack some of your winter clothes for our trip next week. Shanta said the forecast for New England calls for lots of snow and cold. Everyone plans on returning to Boston for a festive Christmas holiday season at the O'Riley's! It'll be so much fun to give Thomas Jr. his first ski lesson on their gently sloping front lawn."

"By the way, did I tell you that Mōr also plans on being at the O'Riley's by Christmas Eve. He doesn't want to miss Colleen's and Millie's special dinner"

"Talking of Boston, I've some more good news. Kim Alosco finished her doctorate in astrophysics and has accepted my old position at Harvard. She'll be quite the prof with her intriguing background! And MJ has invited her to spend Christmas Day with the family."

"I forgot to tell you that Shanta, along with us, is also picking up Thomas in Chicago and Duffy, Millie and Michael in Ireland. Everyone is thrilled with anticipation! It'll be their first opportunity to ride on her command vessel."

"Thomas has already made it known that he insists on sitting in the captain's chair with his namesake on his lap. I pray daily that our son does not acquire his theatric and wacky traits. Something tells me my prayers will not be heard!"

"Hon, it's hard to believe, but this coming Christmas holiday will be Shanta's and Lynkai's longest physical contact with humans. Hopefully, this will prove to be humanity's last moment of resistance, ending all skepticism when it comes to the alien question."

* * * * *

The Presence and Snowflakes now stood ankle deep in the ocean's white bubbly wave wash while continuing to maintain a watchful eye on Ryan.

"Snowflakes, I had lunch the other day with the renowned astrophysicist Dr. Stephen Hawking. I queried him about his thoughts on the causes of Earth's timeless state, but he was circumspect and noncommittal."

"On his arrival here, all his physical disabilities vanished and he was able to converse normally once more. Stephen's personality remains unique; and on occasion, he can be quite aloof. If I had to guess, I think it's partly because he has found the reality of the present to be diametrically opposed to his past stated belief: that *nothing* directs the order of the Universe."

"The good professor is certainly a clairvoyant genius. I have enjoyed our pointed and spirited dialogues over the last months. Particularly when I ask him *who* he thinks I am…or in his world of science… *what* he thinks I am."

"Typically, there is no immediate response. So, I watch him silently for a time as I let him collect his tangled thoughts."

"His eyes betray the fact that he has grappled with, analyzed, and dissected my two questions for far longer than he cares to admit. His ambivalence is born from an inability to overcome certain obstacles, thanks to his long-held anti-pietistic views!"

"But Snowflakes, to my amazement, Stephen recently out-of-the-blue questioned me about a number of Earthly apparitions. Apparitions in places such as Lourdes in France, Fatima in Portugal, La Vang in Vietnam, Zeitoun in Egypt, Guadalupe in Mexico…and the nearly three hundred others that have been authenticated."

"For some reason he assumes that I'm responsible for these apparitions."

"When I ask him who humanity claims it is that appears, he responded without hesitation and with a hint of fervor: 'Why Mary, the mother of Jesus.' In his answer, I detected an echo from the back of his mind, a remembrance of when once he believed!"

Snowflakes threw up her arms in mild protests. "His apparent acceptance of supernatural events, Mary's apparitions, seemingly contradicts his avowed agnosticism. And that makes no sense. He can't have it both ways…accept Mary's extraordinary apparitions and completely reject the idea of a Presence."

"Snowflakes, in most respects you are right."

"I asked Stephen why it is that Mary's appearances on Earth far outnumber those of her son, Jesus?

After a long pause, Stephen thoughtfully replied: "That is quite a mystery. But that is also an unfair, trick question … only God knows the answer to that question… assuming God does, in fact, exist."

I stared at him a long moment with a broad smile.

"Stephen, would it surprise you that I happen to know the answer to that question!

"That's is total nonsense. That implies you are God. You are absolutely not a Deity. I spent the last years of my life debunking the need for a Devine Cause in the creation process or one that holds sway over the universe."

"My dear, dear friend, by now you should realize that life is but a circle, with no beginning and no end. Your being here *now* is proof that your life is part of that circle.

"In time, your lingering doubts will be resolved. While I am not who you may think I am… I am, like you, a part of that life circle."

"You see Stephen, *I AM* the circle!"

* * * * *

Ryan, as he stood drying off after body-surfing, fidgeted a bit as he ran his left hand through his curly jet-black hair that cascaded to his shoulders. He looked up at the Presence, his eyes conveying the paradox that provoked his thoughts.

"I'm afraid to admit that I get more than a little frustrated with those who praise You when You answer their prayers, yet fall silent when they think their prayers are not being answered."

"When these people look in the mirror does their hypocrisy reflect back at them?"

"Ryan, that is a natural reaction to this perpetual dilemma."

"Some would say if a giant tree falls alone in the vast unlimited expanse of a never-ending forest, no one will hear it fall. But I do, and I always will. I will never abandon anyone or anything that I've had a hand in creating!

"And so it is, that I do answer every prayer…but the trick has always been to get that someone to listen when I answer… particularly if they don't like the answer. Remember, humans are perfectly imperfect."

"Snowflake, my dear friend Mōr and you understand that I am the eternal and omnipotent consciousness in the universe. The Cause of all that has been, is, and will be. A living Presence, of energy filled with light and gentle love. Light and love, which like myself, spontaneously came to exist out of nothingness… with an infinite capacity to be."

* * * * *

Merry Christmas to all and as Tiny Tim is fond of saying: "Bless us … everyone." Let it be told here: that the spirit of Christmas is not about seeing, it's about believing!

Believe…that light and love will always be triumphant over darkness and those with an evil bent.

Teardrops fell from Her eye and three identical snowflakes filled with light and love were born, free-falling in the universe… so that *you* might Believe… in miracles!

Made in the USA
Monee, IL
02 November 2026

67384725R00138